Essentials of Short-Range Wireless

For engineers, product designers and technical marketers who need to design a cost-effective, easy-to-use, short-range wireless product that works, this practical guide is a must-have. It explains and compares the major wireless standards – Bluetooth, Wi-Fi, 802.11abgn, ZigBee, 802.15.4 and Bluetooth low energy – enabling you to choose the best standard for your product.

Packed with practical insights based on the author's 10 years' design experience, and highlighting pitfalls and trade-offs in performance and cost, this book will ensure that you get the most out of your chosen standard by teaching you how to tailor it for your specific implementation.

With information on intellectual property rights and licensing, production tests and regulatory approvals, as well as analysis of the market for wireless products, this resource truly provides everything you need to design and implement a successful short-range wireless product.

NICK HUNN is an independent wireless design consultant with over 30 years' experience in design and technical evangelisation within leading-edge technology companies. He has developed products using a range of technologies and wireless standards, including Bluetooth, 802.11, ZigBee and proprietary wireless products, which have achieved numerous awards, including a Queen's Award for Technology. He is Vice Chairman and Executive Director of the Mobile Data Association and former CTO of TDK Systems Europe.

The Cambridge Wireless Essentials Series

Series Editors
WILLIAM WEBB, *Ofcom,* UK
SUDHIR DIXIT, *HP Labs,* India

A series of concise, practical guides for wireless industry professionals.

Martin Cave, Chris Doyle and William Webb, *Essentials of Modern Spectrum Management*
Christopher Haslett, *Essentials of Radio Wave Propagation*
Stephen Wood and Roberto Aiello, *Essentials of UWB*
Christopher Cox, *Essentials of UMTS*
Steve Methley, *Essentials of Wireless Mesh Networking*
Linda Doyle, *Essentials of Cognitive Radio*
Nick Hunn, *Essentials of Short-Range Wireless*

Forthcoming
Amitava Ghosh and Rapeepat Ratasuk,
Essentials of LTE and LTE-A
Barry G. Evans, *Essentials of Satellite Comunications*

For further information on any of these titles, the series itself and ordering information see www.cambridge.org/wirelessessentials.

Essentials of Short-Range Wireless

Nick Hunn
WiFore Consulting

CAMBRIDGE
UNIVERSITY PRESS

CAMBRIDGE
UNIVERSITY PRESS

University Printing House, Cambridge CB2 8BS, United Kingdom

Cambridge University Press is part of the University of Cambridge.

It furthers the University's mission by disseminating knowledge in the pursuit of education, learning and research at the highest international levels of excellence.

www.cambridge.org
Information on this title: www.cambridge.org/9780521760690

© Cambridge University Press 2010

First published 2010

A catalogue record for this publication is available from the British Library

Library of Congress Cataloguing in Publication data
Hunn, Nick.
  Essentials of short-range wireless / Nick Hunn.
    p. cm. – (The Cambridge wireless essentials series)
  Includes bibliographical references and index.
  1. Wireless communication systems. I. Title. II. Series.
  TK5103.2.H86 2010
  621.3845'6–dc22   2010016395

ISBN 978-0-521-76069-0 Hardback

Contents

1   Introduction

Over the past ten years there has been a revolution in the development and acceptance of mobile products. In that period, cellular telephony and consumer electronics have moved from the realm of science fiction to everyday reality. Much of that revolution is unremarkable – we use wireless, in its broadest sense, for TV remote controls, car keyfobs, travel tickets and credit card transactions every day. At the same time, we have increased the number of mobile devices that we carry around with us. However, in many cases the design and function of these and other static products are still constrained by the wired connections that they use to transfer and share data.

Short-range wireless links have the ability to transform the way that these devices share data, whether that is high-speed streaming, or an occasional indication of a change of temperature. To aid this transition, an enormous amount of work has gone into designing wireless standards and the associated chips and software to support them. Despite this, the task of changing from a cable to wireless is still seen as a daunting prospect by many designers; wireless retains its reputation of being close to black magic. The aim of this book is to demystify wireless and remove some of the misconceptions that continue to surround it, as well as to investigate and explain the new mythologies that these wireless standards have generated.

1.1 The growth of standards

In the last 15 years, a number of short-range wireless standards have emerged. Two reasons in particular lie behind this. First is the desire to remove cables from products, driven by the continued growth in portable products. Second, the availability of globally accessible license-free spectrum, along with cost-effective silicon, particularly in the 2.4 GHz and 5.1 GHz bands, has provided an economy of scale that has made the integration of wireless much more affordable.

Not all of these standards have survived – some have disappeared, whilst others have prospered. HomeRF and HiperLAN are largely forgotten. In contrast, Bluetooth and Wi-Fi are present in over a billion devices. With their success, the industry has come to recognise the benefits to be gained from standards, amongst which are the security of supply resulting from multiple silicon suppliers, improved interoperability and robust performance. All of these emerge from competition and a critical mass of engineers refining and evolving the standards.

Behind Bluetooth and Wi-Fi in terms of volume, but targeting specific industrial and automation applications, is an emerging set of standards based on the 802.15.4 radio specification, including ZigBee, 6LoWPAN and WirelessHART. Another newcomer, which offers a new radio optimised for ultra-low power and a future generation of Internet connected devices, is Bluetooth low energy. In addition to these, a wide range of proprietary radios use the same unlicensed bands.

Despite the impressive volumes of chips being shipped, relatively few different applications have come to market. Instead, we have seen implementations dominated by a few 'killer' applications. In the case of Bluetooth, these have been gaming controllers and voice headsets; for Wi-Fi it has been wireless network connectivity. Other standards are still struggling to find their place in the wireless ecosystem.

Two factors affect this: the first is the 'route to volume', which gives a financial advantage to standards that can get a 'free ride' in a mass-volume product. The second is the relative complexity involved in turning any of the standards into an interoperable application.

The free ride is an advantage that has allowed Bluetooth and Wi-Fi to slide down the price curve to the point where they have become cheap enough to enable new applications. To understand this phenomenon, we need to look at the economics of designing and manufacturing silicon chips.

Funding semiconductor start-ups is always a gamble, especially when the companies are aiming to support industry standards. Because several companies will target the same standard, it is a basic fact of life that a significant proportion will fail to gain any share in the market. That is an inevitable consequence of the cost structure for wireless chip designers; the expense of designing a chip means that they need to sell millions to survive. The general consensus is that the cost of designing a complex wireless chipset ranges from around $3 million to $10 million. The higher end of that range typically includes application processors and embedded protocol stacks. That cost can rise even higher if there are multiple radios in the chip. From the other side, manufacturers using these devices need the cost of the chips to be below the $5 price point. In the case of Bluetooth, where over one billion chips are sold each year, the cost is now under $2 per chip. For the highest volume users in the mobile phone industry it is approaching $1.

With these price points and a potential profit of less than a dollar per chip, companies need to sell a minimum of 10 to 20 million chips to cover their development costs. Those that do will make money and gain a significant market share. Those that don't will run out of cash.

For a standard to be able to justify calling itself a standard there need to be at least three manufacturers who are making interoperable products. Without that, there is no guarantee of a real, ongoing, ecosystem. To sustain it requires an annual marketplace of 50 to 100 million chips.

Attaining that volume is difficult. So far, DECT is the only short-range radio standard that has got there by itself. But it managed that by starting off as an expensive, corporate product, with plenty of margin. For others, that's where the 'free ride' comes in. Both Bluetooth and Wi-Fi were built into products – handsets and laptops, respectively – before any general usage occurred. That meant that the volume of chip sales was present to fund the silicon companies and the standards community. It drove the price of chips down so that companies could design and manufacture affordable headsets and access points, which eventually became taken up by consumers, driving a virtuous circle. For these two technologies the chip costs are now sufficiently low that there is a burgeoning market in new product applications. However, without that free ride, they would probably still be relatively low-volume products, if they existed at all. It is an important lesson for standards developers, as well as a cautionary one for designers embracing new standards, who need to be sure that the standard and chips they include in their products will continue to exist for the lifetime of their design.

1.2 Markets

There are some exciting new areas of growth for products using wireless standards. I will cover the market in more detail in the final chapter of this book, but to give an indication of the range of different applications, it is worth pointing out their diversity, both in terms of current areas and those that are poised to emerge in the next few years. Today, many wireless chips in products like laptops and mobile phones are never used. New markets are poised to change that, as wireless will become a fundamental element of these products' functionality.

The first three covered below – games, voice and Internet access, account for over 90% of today's wireless applications. The ones that follow have the potential to join them as major new markets.

1.2.1 Games controllers

The success of the Nintendo Wii and its imitators has provided the largest single use of a wireless standard – in this case Bluetooth. It is a good example of how users accept wireless as an integral part of a product. It's also a good illustration of wireless being a fundamental part of a product's function. Unlike a phone or a laptop, where it is just one of many features, a games controller will not work without it.

1.2.2 Voice

Voice, as exemplified by the wireless headset, is the next largest application for wireless, accounting for the majority of stand-alone Bluetooth chips that have been sold. Riding on the back of an estimated two billion Bluetooth-enabled mobile phones, it is likely to retain that position for a number of years. However, it is believed that only around one-third of the headsets that are sold are used after the first few days.

1.2.3 Internet access

Following close behind is Wi-Fi, which connects laptops and mobile devices to the Internet through access points. Over the coming years we'll see that connectivity extend to other products wanting to report their status to a remote website or monitoring service, making use of the growing infrastructure of public and personal access points. Although usage is increasing, it is estimated that over half of PC-based Wi-Fi chips and over three-quarters of those in mobile phones are never turned on.

Despite the fact that Wi-Fi gives the ability to connect directly to the Web, almost all of its use is by people for email and web browsing. When we look at web connections that do not involve people, the low-power standards of ZigBee and Bluetooth low energy will come to the fore to connect a new generation of battery-powered products to the web, using intermediate devices as gateways.

1.2.4 Internet connected devices

Internet connected devices are key to the growth of a number of emerging markets. The following section provides an overview of these. Chapter 12 explores them in more detail.

1.2.4.1 Health and fitness

Wireless is expected to revolutionise healthcare by linking personal consumer devices to medical services or personal web applications. The area is known variously as telehealth, eHealth and mHealth, the latter generally referring to products that link to, or are connected through, a mobile phone.

The driving force behind this market is the need to reduce costs as the demographics of an aging population puts ever greater pressure on the cost of delivery of healthcare. It also attempts to address personal health management by providing constructive feedback for the third of the population with long-term chronic diseases.

The market is a wide one, spanning sports, fitness and wellness. It also covers the provision of assisted-living devices for the elderly and infirm, which help them to maintain their independence and live at home. As each installation may employ several dozen different simple sensors it opens up a potential market for billions of wireless devices.

1.2.4.2 Smart energy

Smart energy is the remote control of appliances that consume power. Governments around the world are developing strategies to reduce energy consumption and one approach being pursued aggressively is that of controlling how energy is used. Smart-energy initiatives attempt to try to modify user behaviour, either to use less energy, or to spread its use, so that less power generation infrastructure is required.

A key cornerstone of this approach is the supply of smart-energy meters, which inform users of their actual consumption. The next step is for the utility supplier to control appliances around the home to reduce or spread energy usage. Gas and water meters are unlikely to be powered, and wall-mounted and free-standing products, such as thermostats and displays, may need to run on batteries or scavenged power; hence the interest of the ultra-low power wireless standards, including ZigBee and Bluetooth low energy, which are being promoted to address these market needs.

1.2.4.3 Industrial automation

Although a smaller market, there are many high-value applications for wireless within industrial automation and factories, where the cost of wiring sensors is significant. Wireless technology opens up the possibility of installing a greater number of monitoring sensors, particularly on rotating or mobile machinery, or where cabling is expensive or impractical to install.

Much of the cost benefit of wireless in industrial automation is to receive better feedback on the state of machinery, allowing pre-emptive maintenance to reduce downtime and the associated cost.

1.2.4.4 Home automation

Home automation has been slow to take off, but is starting to grow with the availability of wireless products, which are easier to install. The current market is predominantly for alarms, both burglar and safety (smoke and carbon monoxide). Although most of these products use proprietary wireless standards, which are vendor-specific, ZigBee and Bluetooth low energy are developing profiles to address these and bring interoperability to this market. A number of other emerging wireless standards are appearing that specifically address these markets, foremost among which are Z-Wave and the EnOcean Alliance.

1.2.4.5 Consumer accessories

Today, home automation in the form of the TV remote control is one of the most successful wireless applications, albeit infrared. The appearance of wireless in a growing number of products used within the home, plus a desire to extend wireless connectivity to applications on smart mobile phones, is resulting in vendors moving from infrared to standards-based wireless links.

1.3 What is a standard?

It may seem an obvious question, but before moving on, it is worth trying to put a definition around a standard, or at least a wireless standard. Over time, the word's meaning has evolved or been changed, with an increasing number of specifications claiming to be a standard. The following definition is my own view of what a standard is, and helped determine what I have included and excluded from this book.

From a philosophical starting point, the purpose of a standard is to allow devices that adopt it to work together, or share elements of their design. I would argue that a basic requirement is that one must have the ability to implement it using elements of technology from a variety of different manufacturers. In other words, if it is only supported by one supplier of chips and protocol stacks, then it is not a standard. Even if the specification is published in the public domain, if that one manufacturer disappears, then the standard effectively dies. That currently excludes 'standards' like Z-Wave and ANT. They may attract alternative sources of supply in the future, but today they are a single supplier standard, which carries a risk for the product designer. As I've already pointed out, to maintain its viability, a standard needs to ship around 100 million chips per year. Without that, purely financial factors threaten its long-term survival.

The next criterion I apply is that the standard has, or references, a protocol stack that extends up through enough layers to provide the ability to design interoperable applications. If it does not, then it is essentially a building block upon which a standard can be built. So 802.15.4 falls at this hurdle, although it provides the foundation for ZigBee, WirelessHART and 6LoWPAN, while 802.11 scrapes through, at least in its alphabet soup incarnations, through its use of TCP/IP. But it was really the work of the Wi-Fi Alliance that changed its position from being purely a radio and baseband to the status of a proper, interoperable standard. If a standard does not provide this level of definition, then it runs the risk that installation and usability can become poorly defined, making it difficult for users to understand.

My third criterion examines how well the body responsible for the standard ensures that products will work together. That needs

Table 1.1 *State of the wireless standards*

Standard	Application profiles	Multiple suppliers	Qualification program	Enforcement program
Bluetooth	Yes	Yes	Yes	Yes
802.11	n/a	Yes	No	No
Wi-Fi	Yes	Yes	Yes	Yes
802.15.4	n/a	Yes	No	No
ZigBee	Yes	Yes	Yes	Not active
Bluetooth low energy	Yes	Yes	Yes	Yes
WirelessHART	n/a	Yes	No	No
6LoWPAN	n/a	Yes	No	No
Z-Wave	Yes	No	Yes	No
ANT	Yes	No	No[a]	No
Wireless M-Bus	No	Yes	No	No

[a] The ANT qualification is a self-certification.

a qualification scheme, which devices have to pass before they are allowed to market. Without one, designers have too great an opportunity to tinker with the detail of the standard, resulting in products that do not work together. This is where the number of standards really starts to fall.

Finally, I'd add one last requirement, which is that the standard has an enforcement program to allow it to remove non-compliant products from the market. Without this, the qualification process has no real teeth. And the enforcement program must be used. So far, only Bluetooth and Wi-Fi can claim this, although ZigBee has a program in place. Table 1.1 is a snapshot of where the different standards and pretenders are at the moment.

The higher a standard scores in the table, the better the chance that you will find an interoperable ecosystem of products. When a standard can say yes to all four of those points, it can claim to have moved successfully from being a proprietary standard with good PR to being a true standard.

1.4 Choosing a wireless standard

Despite the number of different standards available, and the number of chips being shipped, there are still relatively few different applications that have achieved any volume. A major reason for this lack of diversity has been the comparative difficulty of adapting the standards to support interoperable applications. That is not something that particularly interests silicon suppliers. To reach high volumes, it is generally in the interest of silicon, stack and application providers to concentrate on a few highly focused applications and hope that others will extend the market application areas.

The fact is that most silicon companies have limited resources to support a wide range of different applications. They make their money by selling tens of millions of chips to a few very large customers. To further this cause, they develop reference designs for popular products, such as headsets, access points and PC adaptors. As most products on the market are based on these, there is surprisingly little practical knowledge around in using the standards for other purposes.

Without a comparative understanding of the standards and how to interface with the chips, it is quite difficult both to choose the most appropriate wireless standard and then to design it into an application. Getting to grips with wireless is not easy; designers need to know how to choose a standard, which is an important skill in its own right, how to make connections using it and how to interface it to their data protocols.

In practice, designers can change very little, if anything, within a wireless standard. If they could, it would no longer be a standard. However, most books on wireless concentrate on the fine detail of the one particular standard they choose to address. That can be very interesting, but for most designers it is irrelevant. It is important to know enough to make an informed decision, but as far as a product design is concerned, knowledge of the fine detail of packet format or coding schemes is largely academic.

1.5 Wireless application areas

Although there are many different reasons for wireless, the process for adopting it generally evolves along the same path. The first step is to replace a cable. That may be for purposes of convenience or because of the cost of installing a cable. The latter can be horrific in areas like industrial plants, where laying a cable to a sensor can cost hundreds or thousands of dollars.

Once companies become comfortable with wireless for cable replacement, the next step is an evolution that moves past the point-to-point replacement to the realisation that they can connect a larger number of products. This is where the connection topology diverges from that available with a cable, which invariably imposes a one-to-one relationship, albeit with the option of more complex topology from the addition of networking layers. Although topology adds complexity to the way products connect to each other, at this stage they are still largely designed as discrete entities, with wireless providing extra flexibility in the way they connect.

Finally comes the most interesting stage, where product design starts from the principle that a device is wirelessly connected and the wireless connection becomes an integral part of its existence. Typically, this will imply an automatic Internet connection through some form of gateway. Today, the industry is only at the very first stage of understanding this. Over the next decade, as more designers understand the implication of Internet connected products, it is likely to transform the way products are designed and used.

1.5.1 Standard vs proprietary wireless

The first, and still one of the most common reasons for embracing wireless, is to replace a wire between devices. For many of these applications there is no real need for a standard, as the connection is between two pieces of equipment that are both supplied by a single manufacturer. It is a market where proprietary wireless links are still common.

Where standards become important is when the two ends of the connection may not be made or provided by the same company. Common examples are headsets and mobile phones, or laptops and access points. As soon as the need for a wireless link extends beyond a single company's products, standards offer the interoperability that enables an ecosystem of products from a diverse range of suppliers to interact with each other. We'll keep coming back to the word 'ecosystem'. It is the key aim of a successful standard, enabling a diverse range of companies to make products that can connect to each other's products. The wider that ecosystem, the more one is able to claim that a standard is successful.

The requirements for cable replacement vary with the application. They may be extreme range, (which can be many kilometres), high data rates, dedicated or ad hoc connections. As standards address more applications and expand their repertoire of features, they promote a gradual move away from proprietary radio systems to the use of short-range wireless standards, both because so much of the work is already done within the chip and protocol stack, and also because of the economy of scale that leads to lower-cost silicon.

1.5.2 The importance of topology

One of the first mindset changes that designers and users need to come to terms with when using wireless is that of understanding what their new devices will connect to. With a cable it is very simple – there is a physical connector on each of the two devices that you want to connect and you join them with a cable that has a mating connector.

With wireless, this simplicity of connection disappears, as do plugs and sockets. In theory, a wirelessly enabled device can connect to any number of compatible wireless devices within range.

This presents a problem that needs to be solved to ensure ease of use, but it also introduces a great feature that enables the next stage of product-design evolution. That feature is the fact that wireless standards allow products to have multiple simultaneous connections. This is so different from the paradigm of a cable connection that it usually takes people some time to appreciate its potential. At the everyday level it lets a Wi-Fi access point talk to multiple laptops. It also lets a laptop use a wireless mouse and keyboard at the same time as streaming music to a headset. Or it can allow a large number of sensors to send data to a controller. It opens up new ways of thinking about connectivity, including the duration of each connection allowed by the ad hoc or promiscuous ability of most wireless standards, which allow devices to join and leave the network at will. What that means to the developer is that topology comes into the design equation in a way that does not exist with the wired equivalent. I'll talk a lot about topology throughout the course of this book. It's one of the most important ways to differentiate the different wireless standards. It allows new paradigms that differ significantly from their wired equivalents and enable designers to develop new usage models for their products.

1.5.3 The 'Internet of things'

The connection of devices to the net, regardless of whether by wires or a wireless link is described using the phrase 'machine-to-machine connectivity', more commonly abbreviated as M2M, or known colloquially as the 'Internet of things'.

Connecting devices to the Internet is different from connecting them to each other. In the past, this machine connectivity has been difficult and expensive. New developments in silicon and standards mean that it is becoming both cheaper and easier for wireless devices to connect to the Internet. It can be achieved in a variety of ways. It is currently done using Wi-Fi to connect to an IP address via an access point or Bluetooth to accomplish the same thing by using a handset as a gateway. However, the way that 'things' will connect is fundamentally different from the way that we, as human beings, access the Internet. For these devices, connecting to a web application is a fundamental part of their function. Whether it's a health monitor, an environmental sensor or a home alarm, the associated web application becomes a fundamental part of the product design. Without it, the device has no use.

These products are only just beginning to emerge. Most current standards don't fully address this capability, as they are concerned only with device-to-device connectivity. To achieve the 'Internet of things' these applications require standardisation from device to web application. Emerging standards, such as Bluetooth low energy, 6LoWPAN and the ZigBee Smart Energy v2.0 profile are addressing the need for this device to web standardisation and are likely to become important parts of this new ecosystem.

As connected devices start to appear on the market, they will provide a new challenge for designers, who will need to consider the web applications they connect to as an integral part of the product design. Today almost all the products that we own are unaware of anything else in the world – they have their own function and don't communicate their data – whether that is their condition, what they're doing or what they've measured – with anything else. Because of that, they are designed in isolation; the closest they come to an ecosystem is where there is a common design style applied to a range of products, which is purely physical. As we move forward into an age of connected products, innovative companies will take the opportunity to develop new, disruptive products and service models. Designers contemplating wireless will need to acquaint themselves with cloud-based applications and web services. Although these are outside the scope of this book, references [1] and [2] are good starting points to understand the potential. That entails understanding how application developers, whether an internal resource or an external open-source community, fit into the development cycle. Whichever option is chosen, they need to be an integral part of the product-design cycle, not something tacked on at the end.

1.6 Using this book

With so many specifications and standards to choose from, most people get bogged down in detail before they even get to the basic questions they need to ask.

Wireless standards have more in common than is generally realised. They have evolved from a few basic radio configurations, distinguishing themselves with protocol stacks that adapt these radios to particular connection topologies. Rather than presenting them as competing choices, I will focus on the commonalities, explaining where and why they differ and how these differences affect their use.

To start off, we'll look at the distinguishing features that are relevant to choosing a radio, including range, topology, security, data throughput, latency, robustness to interference and power consumption. These are the core differences you need to understand as the first step in making an informed choice.

Once we've established that, we'll look in more detail at how each of the major short-range wireless standards addresses these points. This book doesn't intend to give you a deep knowledge of the intricacies of the different standards, but enough of the relevant detail to understand the basics of how they work, so that you can ask the right questions of a supplier and understand how to construct an application using the features of that standard. For those who want to delve further, reference will be given to other sources of information and the standards themselves.

After that comes the real meat, where I look at how to move from a standard to a product, by explaining the way in which the radio standards can be used for applications. This looks at the practical side of the wireless parameters I mentioned, including topologies, methods of making connections and securing them, managing power consumption, transmitting different types of data and coexisting with other radios. It also looks at the tools and techniques that will be needed, along with a discussion of the system design and architecture necessary to embed wireless connectivity into a wide range of devices. The aim is to give you the confidence to implement a wireless data link, whether it is a simple cable replacement or a fully Internet connected product.

To complete the design considerations, the book covers the practicalities of approvals, regulatory compliance, production testing and export controls. Although these are rarely discussed, a lack of understanding of these points has the potential to delay the time to market and add tens of thousands of dollars to the development cost. Anyone embarking on a wireless design needs to be aware of these at the start, as they are a key factor in the choice of the most cost-effective standard. I also look at the issue of intellectual property. Each of the standards approaches IP licensing in a different way, which can expose manufacturers to different levels of risk and cost of infringement.

Finally, a closing chapter highlights the major opportunities for short-range wireless, exploring the key markets where it is expected to make its biggest impact.

My hope is that after reading this book, a product designer can make an informed choice of wireless standard, and have the necessary level of understanding to plan the design process and ask the right questions, leading to a successful product.

1.7 References

[1] Michael Miller, *Cloud Computing: Web-Based Applications That Change the Way You Work and Collaborate Online* (Que, 2008).
[2] Gustavo Alonso, Fabio Casati, Harumi Kuno and Vijay Machiraju, *Web Services: Concepts, Architectures and Applications* (Springer, 2003).

2   Fundamentals of short-range wireless

Most wireless standards lie. Or, to be fair, they tend to be quiet about the fact that when they quote their performance or their features, everything is quoted at its best or maximum value. In real life, many of these are mutually exclusive, or are compromised or affected by the environment in which they are used. Range is invariably a trade-off with data throughput. Both of them are trade-offs with power consumption. And so on. In this chapter we'll look at the major parameters of wireless and how they interact with each other.

One important thing to understand is that most of the wireless standards we're covering have more in common than is generally recognised. Each has been optimised to do some particular jobs very well (generally their main application), but all of them can cope with lots of other applications. The problem facing product designers, once they move away from these obvious applications, is deciding which wireless standard best suits their needs.

In this chapter I'll look at the basic features a designer needs to think about in making this choice. It will give you an understanding of the trade-offs that exist and help you to ask the right questions both in your design team and to any supplier of wireless technology. Even if you know which wireless standard you want to use, it lets you start the process of optimising your design, by seeing how the different parameters affect each other.

2.1 Basics

Although there are many things you need to know about wireless, there are three basic prnciples that are really important, as they govern so much about the performance of the connection. They are:

2.1.1 The connection model and topology

That is, the way in which devices discover each other, make connections, maintain connections in power-saving modes and disconnect. There is no real analogy in the cable world, as there you are always connected, but it can become the dominant issue in the wireless world.

2.1.2 Latency, range and throughput

Wireless behaves differently from wires. Data injected at one end don't arrive immediately at the other. Wireless may act as a bottleneck, has variable delays in delivering data and a limited range over which it will work. All of these mean you need to view the transmission of data in a new light.

2.1.3 Security

In the wireless world you need to be sure what you are connected to, as there is no convenient socket to plug into. Once you have worked that out, a wireless standard needs to ensure that no one can overhear or capture the information being sent. The stronger your security, the more it is likely to compromise throughput and ease of connectivity.

Once you've understood these three basics, most of the rest is optimisation and working out the best topologies. Trying to design a wireless network without understanding them will be a painful experience. These are the items that typically differentiate the various standards. In the chapters dedicated to each of the standards I'll try and make these clear. For the rest of this chapter, I'll cover the more common aspects, which generally head up a designer's lists of requirements.

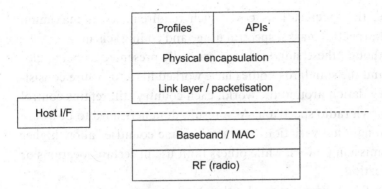

Figure 2.1 Wireless architecture

2.2 Wireless architecture

Before we start, it's worth taking a quick look at the architecture of wireless standards. Most engineers are taught to be familiar with the standard seven-layer OSI network model. Although many people have made valiant attempts to map it onto the actual implementations of wireless standards, it is not a perfect fit. Over the last decade, the architecture of wireless standards has been adjusted to reflect the physical architecture both of the silicon implementations and the physical interfaces of devices that incorporate them, such as phones and PCs. Rather than using OSI layers, wireless standards refer to PHYs, MACs, host stacks and profiles (Fig. 2.1).

Throughout the following chapters, I'll use this format in describing the different standards.

2.2.1 The radio

It is invariably easier to start from the bottom. In the case of wireless standards, that is always taken to be the radio. Wireless standards using the unlicensed bands need to conform to national requirements for usage. These exist to ensure that the spectrum is used fairly. They provide a legal framework, giving regulatory bodies the power to remove non-conforming 'antisocial' radios from sale. All of the standards define radios that conform to these requirements;

hence, the specifications cover such requirements as maximum power, spectral masks, spectral usage and coding schemes.

Although these standards are generally presented as being global, and the standards bodies have worked hard to ensure consistent regulation around the world, each country still retains control of its own radio spectrum. As a result, designers need to be aware of the national variations that exist. Some countries allow higher transmission powers, while others limit use in certain locations or applications.

The RF design is typically the smallest section of any wireless standard. It surprises many designers that the hardware definition may be less than 5% of the total specification in terms of the number of pages. For most designers, it's an area where they have very little ability to change anything, other than the output power.

The other surprise is the way in which the radio affects the performance of the standard. Although the differences in radio specification have some effect, it is nothing like as great as that of the higher-layer stack. This controls the way in which devices can connect and transfer data between nodes and the manner in which the link is controlled. These layers are far more responsible for the different feel of the standards. The radio is just the physical transport.

2.2.2 Baseband: media access control (MAC)

The baseband tells the radio how and when to send data over the air. As well as assembling packets for transmission, it controls access to the spectrum, determining how and when data are sent. That includes working out hopping sequences for frequency-hopping standards, and activity sensing for conditional access standards.

The lower levels of the baseband are responsible for making, managing and maintaining links between wireless devices. I've already pointed out the fact that wireless connections and topology are far more complex than for a wired environment, where this is controlled by plugging in a cable. The mechanisms to manage these links reside within the baseband.

The baseband also controls the lowest level of the security for the connection, both in terms of authentication and encryption. Core security is always located at this layer, as it ensures that security over the air interface is placed as close as possible to the radio transceiver. This allows security to be maintained without the need to wake up a host microprocessor, but may require chip designers to add hardware accelerators to cope with the processing requirements.

At the top of the baseband, most wireless standards define an interface layer. This structure is determined by a number of different requirements:

- A physical requirement, which may be that this is where the wireless chips provide their physical interface. For this reason, it has been standardised in many wireless architectures, allowing an interoperability layer for radio or baseband chips from a variety of different vendors.
- It provides an economical point for PC applications, where all of the higher-protocol stacks can be implemented within the PC. This is true for most Bluetooth and Wi-Fi PC-based implementations.

In some cases, different standards groups may be responsible for the radio and higher-level stack. This is true for Wi-Fi, which uses a radio defined by the IEEE 802.11 group and also for ZigBee, which builds on the IEEE 802.15.4 radio.

This split in architecture means that there is an obvious opportunity for a moderately well defined interface at the top of the baseband. Some standards, like Bluetooth, define this explicitly, even going so far as to specify how it works over the particular physical interfaces of USB, RS-232 and SD. Other standards define a set of programming commands, but leave the physical implementation to the whim of a particular silicon chip supplier.

In the case of Bluetooth products with a USB interface, it is possible to swap implementations from different chip vendors and to expect full interoperability. In contrast, Wi-Fi leaves the MAC interface to a specific implementer, so the same interchangeability does not exist. As well as providing standards-based interface functionality, all chip designers supplement these with proprietary functions that are designed for production test, or proprietary enhancements to the standards. If these are used, designers should be aware that they are committing themselves to that specific vendor.

2.2.3 Higher-layer stacks

The radio and MAC are responsible for making and maintaining the wireless connections and ensuring that data are transported over the air. In most cases they will also take care of at least some of the security and encryption. What they do not do is interact with the applications within the devices. This is the role of the higher-layer stacks.

Higher layer stacks come from a variety of sources. Some, like Wi-Fi, are based on wired protocols, such as TCP/IP. Others are created by organisations that use a pre-existing radio and MAC. Examples of this are ZigBee and many others that use the popular 802.15.4 MAC/PHY. In other cases, organisations like Bluetooth specify both the radio and the higher-layer stack.

Higher-layer stacks provide the link between an application and the wireless connections. They have a vital function in providing interoperability between devices from different manufacturers, as they define standard ways for the data to be packaged and formatted, along with standardised APIs. However, they do not generally know the details of how an application or application area needs to work. For that level of interoperability, standards generally need to add profiles.

2.2.4 Profiles

Profiles are an attempt by standards groups to bring interoperability all the way up to the application level. Profiles arrived after the first generation of DECT cordless digital handsets entered the market. These met the requirements of the standard, but models from different vendors were not interoperable. As a result, customers shunned them and the market stagnated. To address this, the DECT industry developed its generic access profile to provide the additional layer that meant users could mix and match handsets from different manufacturers. It worked and breathed life into the DECT market. It is an approach that has been adopted by many other wireless standards.

A profile basically tells a manufacturer how to use the wireless standard to enable a specific application. In general, pofiles are both prescriptive and complex, as they try to cover the commands, data formats and data protocols used between two devices for a specific application. This may cover how the devices connect to each other, the security requirements that must be met, how to respond and recover from errors and even the functionality of a user interface.

Profiles always contain a number of mandatory features, which must be implemented if they are to be approved. To try to ensure interoperability, standards bodies run qualification programs to which manufacturers must submit their products if they are to claim compliance.

Because of their complexity, there are relatively few profiles across wireless standards, and these have been developed for high-volume applications. Profiles are generally developed by working groups within the standards bodies that have a specific market interest in a particular application.

To address a wider range of applications, we can split profiles into application profiles and transport profiles. The latter exist to provide a means of supporting a wider range of applications. Rather than configuring themselves to a specific application, they are limited to methods of connecting devices, including security and a physical transport interface, which is generally a serial port. They do not detail the way in which data are formatted, so provide little more than the wireless equivalent of an RS-232 cable. However, their flexibility has enabled thousands of different applications to enter the marketplace.

Bluetooth low energy and ZigBee have further evolved the concept of profiles to try and make it easier for companies to develop applications. In the case of ZigBee, this is through the use of the ZigBee cluster library, which provides standardised functions that can be accessed and used by proprietary applications. Bluetooth low energy restricts applications to using a single, well defined protocol, and then uses an object-oriented set of services to provide a simpler, flexible alternative to profiles. Both represent a new-generation approach to the complexity that has grown up within profile definitions.

It should be pointed out that building applications that do not have specific profiles can be a daunting task. A number of module and software manufacturers produce building blocks that contain proprietary APIs to simplify this process. These will be discussed in Chapter 12.

2.3 Wireless parameters

Most aspects of wireless connectivity interact with each other, so it is important to consider all of them when making a choice of standard, or when embarking on a new design. Every application has different requirements, but it is always a good idea to go through the tick list of parameters. We'll look at the key ones that need to be on that list:

- Range,
- Throughput,
- Output power and link budget,
- Interference and coexistence,
- Security – authentication and encryption,
- Power consumption,
- Topology,
- Connection type,
- Latency,
- Usability and commissioning,
- Profiles and interoperability.

Whilst the last two items don't necessarily impinge on performance, they can affect topology and make the difference between a very usable system and one that requires an expert to set it up.

2.3.1 Range

Range is invariably the first question asked by designers when looking at a wireless standard. Because wireless is so intangible, long range seems to provide a first level of comfort. Designers seem to think that if the range is great enough, then everything will work. There is a lot to be said for having more range than you need, because as soon as a radio is brought inside a building, fading, interference and reflections will reduce the range significantly. (Data sheets almost always quote either a 'theoretical' range, which has been calculated or promoted by the marketing members of a standard, or a 'free-space' or 'open-field' range, which is measured in a large empty field on a clear bright day. The latter has the advantage that it's a quantifiable measurement, but neither has much relevance for your real application.)

Range is closely tied up with throughput and output (transmit) power. It is a fascinating subject, which could take up a book by itself. For the sake of brevity, we'll confine ourselves to the major factors involved.

The first question is, 'What is range?' Unlike a cable, which generally either works (when it's plugged in) or doesn't (when it's unplugged), wireless performance degrades as the distance between transmitter and receiver increases. It is not a linear decrease. Generally, the connection maintains a link that supports a data rate close to the maximum the transmitter and receiver can support, until it reaches a point where it starts to decrease, with the fall-off increasing as the separation between them increases (Fig. 2.2). There is no strict definition of where the extreme of range is determined, so different viewpoints can affect the published value.

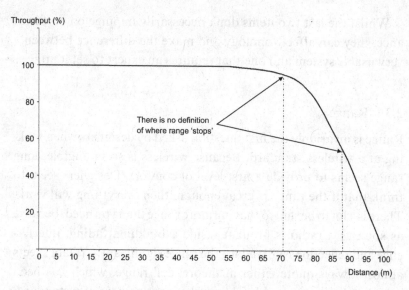

Figure 2.2 Range

There is no agreed definition of where range stops. At one end, it can be considered as the point where throughput falls by 5–10%. At the other, it's where it stops altogether. The point at which the link is lost is not a good way of defining range. Because all of the wireless standards that we're looking at attempt to resend data that have not been acknowledged by the receiver, data will continue to trickle through, even at extreme ranges. There will be a point at which this is so infrequent that a watchdog or link supervision timer will decide that the time has come to abandon the attempt and inform the user that the wireless link has been lost. However, this is at a point that is often far from any usable situation. It will almost certainly have taken the radio into an operating regime where the power consumption has increased significantly. Multiple retries don't just mean reduced throughput, they also mean higher power consumption. The reason is simple, but often overlooked. If you need to send a packet three times, you need to keep the radio powered for a much longer period. It is often a bigger effect than anticipated, as the radio may stay awake while it is waiting to get a chance to resend the message, where it would otherwise have moved to a sleep mode. So running a link at the extreme of its range can have a significant effect on battery life.

For data transmission, it is often taken as the point where the data rate falls to about one-tenth of the maximum, but is still reliable. For voice, it is the point where noise and distortion make the received signal unacceptable. These are both subjective measurements and there is no formal definition of them. Whatever method is chosen to determine range, it will almost certainly be implemented using a different test environment from that experienced by your application, so you should always test range early in a product development. Unless you know that you can control the environment in all of your radio deployments, always aim to work well within the expected range for the target environment, preferably at no more than 30% of the measured range.

What underlies the decrease in throughput is the fact that data start to get lost in the background noise. As transmitter and receiver move further apart, the strength of signal arriving at the receiver gets smaller, until it reaches the point where it is not possible to resolve the signal from the background noise. This is exacerbated in the real world by interference, which may result in individual bits within the data stream being corrupted. The quality of the radio link is expressed by one of the most frequently used measurements within wireless – the bit error rate, or BER.

Bit error rate is a simple, concept, defined as:

$$BER = \frac{\text{number of erroneous bits}}{\text{total number of bits transmitted}}.$$

It is expressed as a number raised to a negative power of ten. Typical values for wireless networks are in the range 1×10^{-5} to 1×10^{-10}. It is not a bounded measurement, or related to a specific time, but it tends to remain constant for a specific physical deployment.

Although the BER is greatly influenced by the physical layout of the radio units, it is also affected by the radio design itself; in

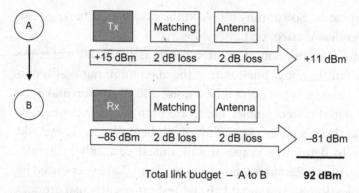

Figure 2.3 Link budget

particular, the ability of the receiver to extract data from background noise. This, in turn, is influenced by the modulation schemes that are used. However, most of this is beyond the control of the user. All that an implementer is likely to be able to affect is the local noise level at the receiver and the choice of receiver chip. This means that similar chips from different vendors may differ significantly in their receive sensitivities. The only way to discover this is to test them.

2.3.1.1 Link budget

Having chosen the radio, there are further parameters that can be changed to affect the overall range. Crucial to this is a concept called the link budget. The link budget is a figure that takes into account all of the features that affect the transport of a packet from one device to another. The two biggest contributors are the output power of the transmitter and the input sensitivity of the receiver. The everyday analogy is that transmit power relates to how loudly you can shout, whereas receive sensitivity is how well you can listen. To be able to support a long-range conversation it is important that the link budget is equally good in each direction.

Both receive sensitivity and transmit power are normally quoted at the RF pins of the transceiver chip (or separate transmitter and receiver packages). To reach the outside world, these pins need to be connected to an antenna. The power quoted is unlikely to be what comes out of the antenna.

To ensure that as much of the energy from the transmitter reaches the antenna rather than being reflected back and lost, RF components are designed to have impedances of 50 Ω, so that a perfect match can be obtained through each circuit element. Unfortunately, manufacturing tolerances mean that they are not perfect. At gigahertz frequencies, the wavelength is just a few tens of centimetres. This is the same order of magnitude as the size of a printed circuit board, and tracks begin to behave more like components than pieces of wire. Putting these together gives a situation where matches are less than ideal. To rectify this, designers use matching networks, typically comprising capacitors, resistors and inductors, to try to restore a perfect match.

These components inevitably result in some of the signal being lost, as the components have an insertion loss, which loses some of the signal. This is effectively an impedance that 'eats up' some of the signal. Perfect components have no insertion loss. Real ones consume anything from 0.5 dB to several dB. If a single antenna is shared between receiver and transmitter, the design will need an RF switch, which also eats up some more of the link budget in both the transmit and receive chain. In most designs, imperfect matching also causes a percentage of the signal to be reflected back to or from the antenna. As a result, what comes out of the antenna is almost always less than what came out of the chip (Fig. 2.3).

The radio's transmit power and receive sensitivity are defined using the logarithmic decibel scale, where:

$$\text{Power(dB)} = \log 10 \left[\frac{\text{power1}}{\text{power2}} \right].$$

Most measurements are expressed in dBm, which is the measured power compared with a reference of 1 mW. Table 2.1 gives the common levels for short-range wireless devices.

Table 2.1 *Common dBm values*

Power (mW)	dBm
0.1	−10
1	0
4	6
10	10
100	20
1000 (1 W)	30

For antennae, the power ratio refers to the gain of the antenna. Antennae have a gain or, in the case of small ceramic or printed circuit antennae, a loss. This needs to be added to the calculation of the link budget as it affects the amount of the signal that gets either out of or into the radio chips. In the case of directional antennae, these values can be major components of the link budget.

A number of different units are used, depending on the directional characteristics of the antenna. The most common is dBi, which refers to the isotropic gain, assuming that it is equal in all directions. A less common usage is dBd, which compares the gain with that of a half-wave dipole.

Regulators tend to use a figure known as the equivalent isotropically radiated power (EIRP), which is the amount of power a device would radiate if it were using a perfectly isotropic antenna. It is referenced to a perfect isotropic antenna with a 1 mW output. In the USA, an EIRP of up to 30 dBm is allowed in the 2.4 GHz band. Elsewhere in the world, it is generally limited to +20 dBm, and in some countries to +10 dBm.

Radios are also given classes, which define the transmit power range. This is most common with Bluetooth radios, where the classes are defined in Table 2.2.

The final use of dBm is for receive sensitivity, where it refers to the minimum signal level that the receiver can resolve. Typical receive sensitivities are in the range of −110 dBm to −70 dBm (Table 2.3).

Table 2.2 *Bluetooth output classes*

Class	Maximum transmit power (mW)	Maximum transmit power (dBm)
1	100 mW	20 dBm
2	4 mW	6 dBm
3	1 mW	0 dBm

Table 2.3 *Common receive sensitivities*

Minimum detected signal (pW)	dBm
0.01	−110
0.1	−100
1	−90
10	−80
100	−70

With the terminology sorted out, we can come back to the link budget. In many cases this link is not symmetrical. It's not uncommon for a central device to connect to a lower-powered peripheral. As the wireless protocols we're discussing employ handshakes, where control messages or data are transferred in both directions, the link-budget calculation needs to be performed in both directions (Fig. 2.4).

If the calculations are different for the two directions, the lower figure will be the limiting aspect of the connection. The important point that this illustrates is that there is no real benefit in increasing the power or sensitivity of just one side of the link. If both the output power and the receive sensitivity of one unit are improved, then there will be an improvement. If only the output power is increased, then it is unlikely that the cost will be justified.

Range can obviously be improved by employing directional antennae, and demonstrations achieving connections over many kilometres have been made with most of these standards.[1, 2, 3]

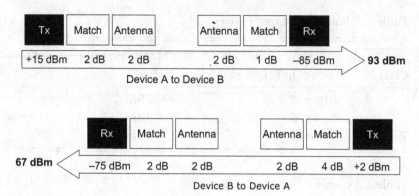

Figure 2.4 Asymmetry of link budget

These shouldn't be confused with general-purpose range measurements, which should be made with antennae that are essentially omnidirectional. Adding directionality is a decision that depends on the application and is only really suitable for cases where the antennae can be fixed in known orientations. That may be acceptable for fixed devices, but it is unlikely to be acceptable for mobile ones, where the antenna orientation is constantly changing or unknown. You will only get a basis for comparison by running tests with your intended antenna.

Knowing the link budget allows a designer to start to understand the range of a system. The link budget can be used with a number of different theoretical models to predict the operating range. A general rule of thumb is that increasing link budget by 12 dB will result in a doubling of range. However, in practice, the local environment will exert a large qualifying effect on the range. Despite this, link budget does still remain the best indication of how a system will perform and is an invaluable indication of the effect of changing components at every point of the radio chain. It is also a good indication of asymmetry within a radio link.

2.3.1.2 Other factors affecting range

Implementations demonstrating extreme ranges usually resort to directional antennae. Designers pursuing this route should be

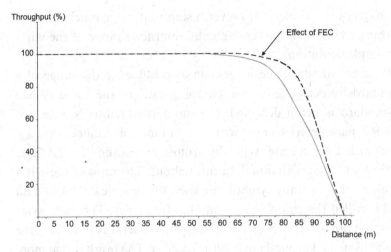

Figure 2.5 Effect of FEC on range

aware that increasing output power or using directional antennae may move a wireless solution outside the limits of what is allowed in a particular country. As a general rule, the more a designer pushes the limits of a wireless technology, the less likely it is to be legal in a majority of countries.

A number of other techniques may be employed by standards designers to tweak the range. One of the factors determining the range is the coding of the radio signal and how data are protected. As range is a measure of the point at which the ratio of bad to good packets becomes unacceptable, increasing the chance that a packet can be repaired gives a real improvement to the link budget. It does not extend the range, but improves the throughput as the radio is used close to the extreme of its range. Adding forward error correction to a packet can be equivalent to adding between 3 dB and 6 dB to the link budget, as illustrated in Fig. 2.5. A similar effect can be seen with retries on synchronous (voice) links.

If greater range is required, there are alternatives to trying to improve the link budget. An installation can use repeaters or routers to increase the number of steps for a packet to get from sender to receiver. This is inherent in mesh standards like ZigBee. It can also be provided by routing over a backbone network, which may or may not be wireless. However, using routers or repeaters does nothing to improve the fundamental internode range of the wireless implementation.

Wireless standards designers can also influence the range of a standard by deciding on the 'chipping rate' of the radio. When a standard is being designed, the standards architects have two key RF parameters to play with: the number of symbols that are transmitted per second (typically around one million for 2.4 GHz radios) and the maximum data throughput. The ratio of these is a measure of how many symbols are used to convey each bit of data and is called the chipping rate, or chip frequency. The more symbols that are used for each bit of data, the easier it is to resolve the signal from background noise at the receiver. The result is that more sensitive receivers can be produced, so the link budget is increased. As always, nothing comes for free and here the price demanded is lower throughput or, if the symbol rate is raised, higher power consumption. A similar effect can be achieved by increasing the modulation index of the transmitter. These are second-order effects and are specified and set in stone by the standard. A designer is unable to influence them, but it is useful to understand them when comparing different standards.

One area where a designer may have some ability to improve range is in the use of multiple antennae. Some radio chips allow multiple receive antennae to be used, automatically selecting the one with the stronger signal. This technique is used in earnest in 802.11n, which implements multiple input and output streams (MIMO).

2.3.2 Throughput

As we've seen, throughput falls as the range increases and the BER rises. For standards that support connection-oriented data, this can be further complicated by the baseband trying to optimise the link where different coding schemes are supported.

To get the highest throughputs, wireless standards attempt to cram more than one bit of information into each bit transmitted.

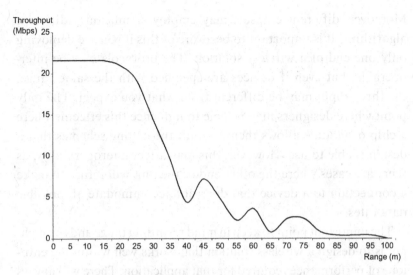

Figure 2.6 Range vs throughput for 802.11g

Both Bluetooth and Wi-Fi do this, and we'll cover the techniques in more detail in the respective chapters. They maximise throughput by trying to select the most complex coding scheme that the link quality will support. As the link quality declines, they will automatically step down to a less aggressive coding scheme until they achieve a reliable link.

It is fairly obvious that the more aggressive the coding, i.e., the more data crammed into each bit, the more susceptible it will be to noise, so the highest data rates will typically have the lowest range. Coding is a step function, not linear, so the renegotiation schemes that are used end up producing range versus throughput plots that are far from intuitive. Figure 2.6 gives an example of spot throughput figures for an 802.11g link. This is made complex by the fact that a number of fundamentally different coding schemes are dynamically negotiated as the two devices are moved apart, each of which has different sensitivities to BER. (Throughput is normally measured in bits per second, or megabits per second for the faster radio standards.)

The designer has no access to the algorithms that select the point at which basebands will make the decision to change coding.

Moreover, different chipsets may employ significantly different algorithms. It is important to be aware of this if you are deploying only one end of a wireless solution. The protocols may be interoperable, but even if devices are specified with the same range, the throughput may be different from what you expect. The only point where designers may be able to influence this effect is where a chip or module allows them to limit the coding schemes that a design is able to use. However, this may affect interoperability, as there are cases where the other end of the link will refuse to make a connection to a device that does not accommodate all possible data rates.

The important point to keep in mind regarding range and throughput is to design a wireless solution that works well within the envelope of performance required for that application. There will always be locations where the application is deployed that are significantly worse than the test sites, so make sure that there is plenty of overhead. But as we will see, that introduces other trade-offs.

2.3.3 Interference and coexistence

Short-range wireless systems always work really well in the test lab. In the real world they have to cope with the fact that they need to share the spectrum with other radios. Spectrum is a finite resource and the license-free bands are regulated to try and ensure that a wide range of different devices have a good chance of coexisting. Having said this, when the regulations were formulated, the regulators probably had no idea that so many different devices would be sold that would attempt to use the spectrum.

Spectrum is something that is not intuitive to many electronics designers, who have grown up with Moore's law, where everything evolves to go faster. In contrast, spectrum is better thought of as a road network. When there are few vehicles, they can travel at their fastest speed. As the numbers increase, the overall traffic speed decreases and if the volumes increase even more, or there are accidents, the whole system can grind to a halt. The analogy of numbers of vehicles with numbers of wireless devices works well. Accidents can be considered analogous to interference between different devices.

If two radios within range of each other both transmit at the same time and at the same frequency, this results in signals arriving at the respective receivers that interfere and are likely to be corrupted. Radios using reliable data links will not get an acknowledgement of reception and so they will attempt to retransmit the data. Different radio standards use different techniques to try and ensure that they don't clash with each other on the retransmission, but even if the next transmissions do not overlap and are successful, it means that the throughput will have decreased as a result of retransmission. And in a situation where this starts to occur, it will probably happen repeatedly.

The designers of the different standards have added features to cope with this interference, but in most cases these are aimed predominantly at ways of mitigating interference with other radios using the same standard. Problems can multiply where there are several different radio standards operating within range of each other, as their individual mitigation schemes can work against them, making the situation even worse.

To understand this, we need to look at the way in which different radios use the 2.4 GHz spectrum. In this chapter we'll just compare Bluetooth and 802.11, as they use the spectrum in totally different ways. They also both support streaming applications that can consume a lot of the spectrum. In contrast, Bluetooth low energy and ZigBee generally transmit infrequently and do not cause interference problems, although they may suffer from the transmissions of more aggressive radios.

Each standard defines a spectral mask for its transmission, which is a set of limits on the output power within and around each transmit channel. These are defined to meet the requirements of global regulatory regimes as well as the practical performance of low-cost radios. In an ideal world, a radio's transmissions would reduce to zero outside the extent of its channel. Real radio implementations are not that perfect; they provide a peak output at the centre of the channel, and the power then falls off gradually at either side. Obviously, it is important for it to fall as quickly as possible, so that it does not interfere with a radio working on an adjacent channel, but to do that requires elaborate radio designs and complex filters, which are expensive. As a result, each standard specifies a set of pragmatic limits which radios supporting that standard must comply with.

As well as the issue of not interfering with a radio on an adjacent channel, there is generally a tighter requirement at the two ends of the band, where spectrum is allocated for other applications. To help protect them from each other, most standards incorporate guard bands at either end of the allowed spectrum, which, although they lie within the allowed license free band, are set aside to ensure that transmissions do not extend beyond the band edge.

Going back to the characteristic of the power output, radio transmitters have spurious emissions, which are transmissions that fall outside their channel. In general, the more complex the coding scheme that a radio uses (which normally corresponds to a higher data rate), the more spurious emissions there are as you approach the edge of the channel. These mean that 'faster' radios are the most problematic for designers. If high output powers are required, then extensive use of filters is likely to be necessary.

In the case of the 2.4 GHz band, in most of the world there is a tight requirement to limit spurious emission at the top of the band, as one of the next users of spectrum is satellite telecommunications, which employ particularly sensitive receivers to pick up weak satellite signals. To see where problems arise, we need to look at the way the different standards use the spectrum.

802.11 radios transmit at a set frequency, with a bandwidth of 22 MHz (Fig. 2.7). The 2.4 GHz band is divided into 14 of these channels, spaced 5 MHz apart, and, depending on the country in which they are deployed, radios are allowed to operate in some or all of these.

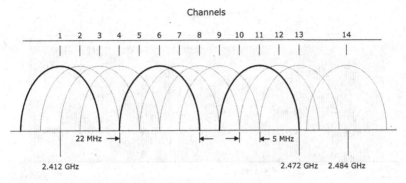

Figure 2.7 Spectrum usage for 802.11b and g

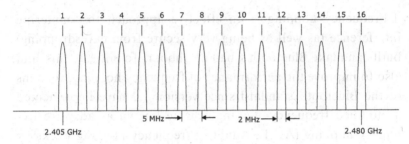

Figure 2.8 Frequency bands for 802.15.4 and ZigBee

802.15.4, the radio that is also used by ZigBee, takes a similar approach, but uses much narrower channels, only 2 MHz wide, spaced every 5 MHz (Fig. 2.8).

Bluetooth takes a different approach, using frequency hopping, dividing the same spectrum up into 79 different channels, each 1 MHz wide (Fig. 2.9). The radios jump from channel to channel in synchronisation with each other 1600 times every second. The reasoning behind this frequency hopping is that if a Bluetooth radio encounters interference and a packet is lost, then it will be repeated at a different frequency, where there is a good chance that there is no interference. However, frequency hopping means that devices need to have a common notion of time. That implies that each device must have an accurate clock, which has an impact on cost and power consumption.

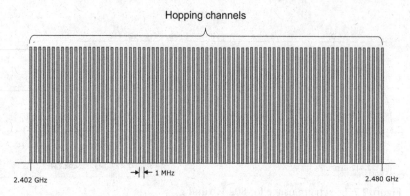

Figure 2.9 Spectrum usage for Bluetooth

Frequency hopping is an effective approach to countering interference (as well as being very secure from eavesdropping) but it can make Bluetooth a bad neighbour. To mitigate this, and also to improve the performance of frequency hopping, versions of the Bluetooth standard since version 1.2 have implemented a modified frequency-hopping scheme known as adaptive frequency hopping (AFH). Adaptive frequency hopping works by scanning the spectrum, looking for channels which are being used. The radios then modify their hopping sequence to avoid these channels (Fig. 2.10). That benefits the hopper, as it is less likely to transmit on a channel that is also in use by another radio and it benefits radios like 802.11 and ZigBee, which sit on a single channel, because Bluetooth avoids transmitting on the channels where they are operating.

Interference between radios is normally uncoordinated, i.e., the radios have no real knowledge of what other radios are planning to do. Other than by using adaptive frequency hopping they can do little else to mitigate the chance of interference. There is a more special case, known as colocation, where two different radios are implemented within the same device. In this case it is possible for the two radio basebands to communicate with each other to schedule their transmissions so that they minimise their mutual interference. We'll look at that case in Chapter 4, where we see how

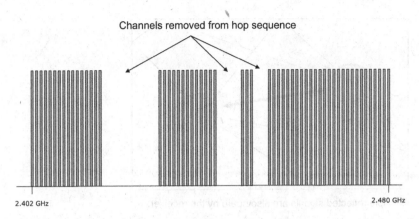

Figure 2.10 Using adaptive frequency hopping (AFH)

the Bluetooth version 3.0 standard copes with using both Bluetooth and 802.11.

As well as interference between radios, reflections of a radio signal can also cause problems. This is known as fading or multipath.

2.3.3.1 Multipath fading

Although simulations can give an understanding of the issues of interference, both with other radios and the surrounding environment, there is only one sure test, which is to try it in the actual location. Real life has a habit of confounding the best of radio designs, whether that's because of natural effects like seawater, or interior designers who think that metal-clad walls look sexy.

These real-world effects are exhibited in the form of multipath fading, also known as path loss. Multipath fading occurs because reflective elements in the environment result in a multitude of different signal paths between the receiver and the transmitter (Fig. 2.11). When the reflected signals start to become a significant proportion of the main signal, then constructive or destructive interference will occur. Destructive interference will increase the BER, reducing the link budget and the range. For narrowband signals in the 2.4 GHz band, path loss can be as great as 30 dB.

For the radio designer, the best approach is to understand the problem and then try to ensure that the design is working well

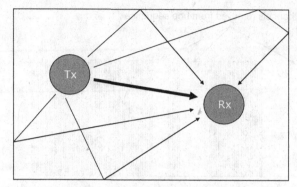

Multiple reflected signals are also heard by the receiver.

Figure 2.11 Multipath fading

within its maximum range, so that there is plenty of headroom. Frequency-hopping radios tend to be more resistant to multipath fading, as the path length at different frequencies will differ, giving them a better chance of eventually getting data through.

2.3.4 Topology

Cables have topology only in as far as they can be connected via different types of hubs or switches. In contrast, wireless networks allow invisible connections between a variety of different devices, in some cases at the same time. The manner in which these connections are allowed between devices is called the network topology.

Topology is often ignored when choosing a radio standard, despite the fact that it is one of the major differentiators between them. Topology becomes important as soon as we get rid of cables. Cables have some inherent capabilities that we tend to forget because they're so obvious. One is that the plugs and sockets that come with cables define exactly which two products they are connecting. Another is that the connection tends to be secure, as physical access to the cable is normally limited by its location. Radio removes these advantages and as a result has to work hard to attempt to replicate or improve on them.

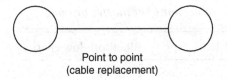

Point to point
(cable replacement)

Figure 2.12 Topologies –
point to point

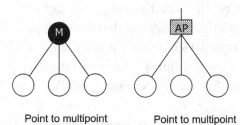

Point to multipoint
(piconet)

Point to multipoint
(client server)

Figure 2.13 Topologies –
point to multipoint

Before looking at how wireless standards cope with association – the process whereby they are configured to connect with each other, we'll look at the evolution of different topologies that become possible once we take the step of removing the cable.

The first, most straightforward and still most widely used is simple cable replacement, which is also what that particular topology is called (Fig. 2.12). It's the exact analogue of a cable, but with the removal of the physical constraint of a piece of wire. All of the wireless standards can perform it, differing predominantly in their speed of throughput.

Next comes point-to-multipoint topology or the piconet (Fig. 2.13), where one device acts as a master or central device, connecting to several other devices. This is also known as a star network. The master may be able to transmit to all of the other devices at once using broadcast messages or, more normally, use timeslots to hold individual conversations with each of the peripheral devices in turn. Different standards allow differing numbers of concurrent live connections in a piconet. They may also allow you to specify the level of quality of service for different connections.

Table 2.4 *Maximum number of connections per master or node*

Bluetooth	802.11	802.15.4	Bluetooth low energy
7	255	20	2 billion

Broadcast is the term given to the mode where the same message is sent to all other devices. Broadcast messages are not normally acknowledged, because of the likelihood that simultaneous acknowledgements from multiple devices would interfere with each other. Think of broadcast as a clock chiming or someone shouting, 'Fire!' Directed transmissions to a specific receiver are called unicast transmissions.

With practical implementations, physical constraints rapidly start to come into play. The maximum number of connections is determined by the size of the local address field within each standard, as shown in Table 2.4.

In practice, there are limitations which mean that these numbers are rarely achieved or even approached. Outside broadcast mode, connections need to be time multiplexed, with individual nodes being addressed one after another. The first limitation to increasing the number of connected devices is that of the memory required to hold all of the addressing and connection information for multiple connections. The second is that bandwidth is shared between devices, so the individual data throughput rapidly diverges from the headline figure. If more than a few tens of connections are made, individual data rates for each connection plummet to a few kilobytes per second. This may not be a problem for a sensor network that only requires a reading to be sent every few seconds, but is inadequate for most other applications.

These practical considerations typically constrain the number of connections per node to a fraction of the maximum possible. For low-cost, single-chip implementations it is frequently well short of the maximum advertised number and can be as low as three for Bluetooth and ten for other standards. Increasing the number of connections invariably needs additional processor and memory capacity within the master node.

Although topologically identical, I consider client server as a distinct form of piconet or star-network topology, as it implies a central unit that acts as an access point to provide connectivity to another network. Another reason for making this distinction is that the security and association management are often different between the two configurations of client server and piconet.

This is the architecture that has made 802.11 so successful; where an access point allows multiple clients to connect to a separate network. 802.11 can also operate in an ad hoc mode, in the same manner as the piconet shown in Fig. 2.13, but this is less well developed as a standard and not currently embraced by the Wi-Fi Alliance as part of the Wi-Fi standard. The Wi-Fi Alliance has announced a new standard to support ad hoc connections, called Wi-Fi Direct, which should be available in the second half of 2010.

A limitation of the piconet is that a slave is a slave. It can only communicate with its master device. If it wants to talk to another device it must remove itself from the first network and make a separate connection to the second network. If slave nodes are given the ability to talk directly to each other, the topology develops into a cluster network (Fig. 2.14). Cluster networks have the advantage that some of the workload can be removed from the master, as transactions no longer need to pass through it. However, the master needs to set up the routing tables in the cluster nodes in the first place, which generally increases its complexity. Clusters also require slave nodes to have more intelligence, to manage their state of awareness for incoming messages. Clusters are not commonly found by themselves. They are an important conceptual step towards full mesh networks (q.v.).

The next level of complexity comes with the scatternet (Fig. 2.15). This is an extension of a piconet, but allows a slave or peripheral device to be simultaneously connected to two (or potentially more) networks. It's a topology that is little used. It exists within the Bluetooth standard, but remains largely academic. It

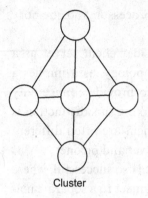

Figure 2.14 Topologies – cluster network

Cluster

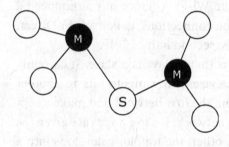

Figure 2.15 Topologies – scatternet

Scatternet

has the limitation that although a device may be part of multiple networks, that device generally only functions within one at a time – it is not able to act as a bridge between the two networks. For that to occur, we need to move to the tree or hierarchical network (Fig. 2.16).

At first sight, tree networks look exactly the same as scatternets. The difference is that the backbone nodes of a tree network have routing capabilities, whereas the connecting nodes in a scatternet may only share data. This means that any node connected to the backbone can send data to any other node within the tree. As the network becomes more complex, the nodes forming the main spine

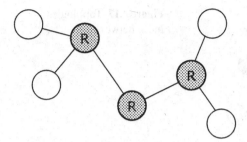

Figure 2.16 Topologies –
tree or hierarchical network

Tree network

require an increasing amount of memory and processing power
to manage the routing of packets. So although tree networks (and
their big brother – mesh networks) are often employed for low-
power sensor networks, there may be a requirement for the routing
nodes to have a substantial power supply. Within these networks,
there is normally one node, usually called the coordinator, which is
responsible for configuring the entire network. Tree networks can
be made up of joined cluster networks, which help to take some
local processing away for the routing nodes. These are called cluster tree networks and move us towards the ultimate topology of
mesh networks.

At the top end of complexity is the mesh network (Fig. 2.17),
where individual nodes may be able to talk directly to each other
if they are within wireless range, or through a selection of alternative routes. Whereas the tree network uses a single backbone for all
inter-cluster routing, meshes allow multiple routes throughout the
network. This adds redundancy to the network, allowing traffic to
be maintained even if some of the nodes or links between nodes
fail. ZigBee is the only wireless standard that currently offers this
topology.

As a general rule, the more complex the topology, the more complex the technology needed to support it and the more complex
the installation and commissioning. With wireless it is advisable
to go for the simplest choice that meets the needs, trying to take

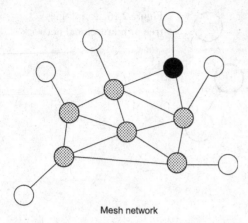

Figure 2.17 Topologies – mesh network

Mesh network

into account future expansion requirements. Obviously that will be tempered by other factors, such as range, robustness and speed, but there is little to be gained, and often much to be lost by choosing complexity for its own sake.

2.3.5 Security – authentication and encryption

As mentioned above, cables have the advantage of security. You know what they are attached to and there is limited opportunity to interfere with the data being passed down them. With wireless it is a totally different situation.

Devices connecting through a wireless link need to be sure that they have connected to the correct device. As anyone with a suitable receiver can listen in to the wireless conversation, they also need to encrypt their data, so that an eavesdropper who intercepts it cannot decode it. These two processes are known as authentication and encryption.

Cracking wireless security is a popular target for researchers wanting to make their name. They provide a good test of the level of security of a wireless standard and most standards have suffered as a result, learning the hard way that security is not easy. Both Bluetooth and Wi-Fi have had to evolve to more complex security schemes as previous ones have been found lacking. Other standards with a smaller market share, or less exposure to the consumer market, have received less attention from the hacker community and may yet need to reassess their implementations. It is likely that all of them will release further enhancements as new means of cracking security procedures evolve.

Each of the standards covered in this book has a credible approach to security. If you implement the latest recommendations, then your product should be secure. For those who want to delve further into the subject, the next chapter digs down into the detail of the different approaches each has taken to security.

2.3.6 Power consumption

Many wireless products are designed to be mobile, which implies that they will run on batteries. Many static products, like light switches and energy meters have a requirement to be maintenance free and need to run off a small battery for many years or tens of years. As radios, sensors and application processors become even lower power, these will be joined by a new generation of products so low powered that they will run on scavenged energy. For all of these, minimising the power consumption of the wireless link is critical.

It is important to be clear about the distinction between a low-power radio and a low-power application. The two are often confused. A low-power radio enables data to be sent efficiently over the air. It includes power-management schemes to ensure that the radio can remain in deep sleep modes whenever possible.

A low-power application encompasses the whole device. Not only does the rest of the circuitry need to be low power, but the application needs to limit the amount of data sent between devices. However clever the design of a radio, it requires a finite amount of energy to transfer data over a wireless link. If too much data are being transferred, then however good the radio design, the battery will not support the life that the application demands.

As we saw with the discussion of range and throughput, power consumption goes down as the range and throughput decreases. There's a complex relationship between coding schemes and energy efficiency. As more bits are squeezed into each packet, the processing requirements increase, but so does the BER for the link, so although a higher coding rate may be technically more efficient, as it requires fewer packets, this may be offset by the need for a higher transmit power to get a reliable transmission. As is always the case in wireless, there are no easy trade-offs. For high-throughput applications life is even more difficult, as there is energy associated with each bit of information transmitted over the air. Here, UWB has the promise of significantly lower power consumption, but is still very early in its development. It is unclear when global UWB standards will settle down.

Most low-power or ultra-low power applications are centred around sensors, which only need to communicate their state intermittently. This may be on a timed basis, or as the result of an event, as is the case with a fall alarm or a thermostat. Here a different set of factors comes into play. Rather than looking at the energy per bit of transmitted information, the battery life depends on the duty cycle and how quickly a radio can move from its sleep state to perform a wireless transaction and then return to sleep.

This is the usage case addressed by many 802.15.4 radio-based standards, including ZigBee, as well as Bluetooth low energy. There are also a large number of proprietary and sector-specific standards that perform extreme optimisations to address the lowest power needs. These include ANT, [4] Z-Wave [5] and the EnOcean Alliance.[6]

In this ultra-low power world, there are two important parameters (Fig. 2.18). The first is the sleep current. Ultra-low power sensors spend most of their life asleep, waking up as a result of an external event or an internal timer. When they are sleeping, they need to consume negligible current, ideally no more than the leakage current of the source powering them. Most manage deep-sleep

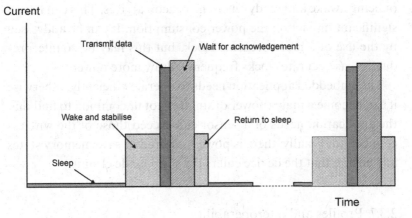

Figure 2.18 Duty cycle

currents of a few microamps. Some of the best low-power radio chips can survive on a few hundred nanoamps.

The second parameter is how quickly the device can wake up, assemble and transmit the data packet, wait for an acknowledgement (assuming that the protocol requires one) and then return to the sleep mode. The length of this time is principally determined by the standard, which specifies the number and size of packets that need to be sent over the wireless link. Standards that aim for this ultra-low power market work hard to minimise the need for unnecessary transactions, and can give total 'ON' times less than 5 milliseconds.

There are a couple of points to note, which are illustrated in Fig. 2.18. The first, counterintuitive, one is that for many radios the current consumption in receive mode is greater than it is in transmit mode. So if a transmission needs to be acknowledged, it is important that it is done as quickly as possible, so that the receiver does not sit around burning current. The second point is that there are times associated with the device waking up, stabilising and performing its measurement before it can transmit.

If a device needs to wake up to listen to a beacon from a master node, then the inevitable uncertainties in the clock accuracy may mean that it needs to wake several milliseconds earlier to be sure of being awake and ready when the beacon arrives. This can have a significant impact on the power consumption. It can be addressed by the use of a more accurate clock, but this may be counterproductive, as accurate clocks frequently draw more power.

The embedded application needs to operate efficiently; otherwise it can become a major power drain. It is not uncommon to find that the application needs of a sensor can exceed those of the wireless connection. Finally, there is power required to save memory states and ensure that the device enters the sleep mode cleanly.

2.3.7 Profiles and interoperability

The reason for choosing a standard is often to gain interoperability with other products from different manufacturers. Bluetooth and Wi-Fi are good examples, where product manufacturers expect their products to work with other products already on the market.

Application-level interoperability is not generally part of a core standard. The core limits itself to describing the protocol used to connect and transfer data. If an application is defined, it is generally through the use of a profile or service class that sits on top of the core. These describe the behaviour and mandatory requirements for specific common applications. Bluetooth, ZigBee and Bluetooth low energy rely on them to give a baseline for interoperability. The Wi-Fi Alliance has fewer options, as it currently only supports an infrastructure mode of connecting client devices to an access point. However, it still includes extensions for advanced power management, initial set-up and multimedia which are not dissimilar to profiles.

Profiles are often considered complex by newcomers to wireless standards, as they move away from the pure protocol definition and begin to encompass application functionality. In the case of Bluetooth and ZigBee, there is the complexity of several different profiles that appear to address similar applications. If you want the best interoperability, it is always worth finding out which profiles are being supported by the industry at large, by looking at the relevant public certification databases for Bluetooth and ZigBee.

What a profile does is describe the behaviour to enable an inter-operable application. Generally this defines which specific transports within the underlying specification are used; how they are negotiated and set up; associated data control channels and commands; error routines and messages; and security requirements. Profiles rarely extend into the application itself, although those that are targeted at devices with minimal user interfaces, such as headsets, may do so. This means that in most cases, manufacturers are still able to differentiate their products above the profile interoperability level.

2.3.8 Voice and latency (quality of service and synchronous transmission)

One other key way in which a wireless connection differs from a cable is that there is a finite delay between data being presented at the transmitting end and being received at the other. Although cables exhibit a small transmission delay it can normally be measured in nanoseconds and is fixed for a particular cable. With wireless it is at least several milliseconds and can be several seconds. This is another of the counterintuitive aspects of wireless that designers need to be aware of.

Latency, which is the measure of this delay, is different for each different wireless standard. It is also implementation dependent, and may vary for different silicon chips. In Chapter 9 we'll look at how it can be addressed in real designs.

Having said that, voice, which is one of the applications most susceptible to latency variations, remains one of the most widely used applications for short-range wireless, in the form of Bluetooth headsets. Voice imposes a special requirement on the data link, which is that it must most closely emulate a cable in giving near instantaneous transmission.

For that reason, voice is very different from either data transmission or audio (music) transmission. Normal wireless data transmission starts with the premise that some packets will be lost, so behaves like an IP transport, where packets are acknowledged and resent. This poses a problem for voice transmission, as delay in a live conversation is obvious and a very undesirable artefact. Hence Bluetooth, and other standards which transport voice, do so without any significant digital processing, but use a simple analogue-to-digital conversion to allow 'live' transmission of speech.

This is called synchronous (SCO) transmission. Specific time-slots are set aside for packets to be transmitted, and the norm is to accept that if a packet is not present, it is discarded. At the receiver, a variety of strategies can be used when a packet fails to appear. Either the previous packet's audio content is repeated, or the device can revert to replacing the missing packet with 'comfort noise' or silence. More recent versions of Bluetooth have added a limited number of packet retries for synchronous transmission: this is known as extended SCO or eSCO. This can help in noisy RF environments, as well as by extending the effective range of a voice link.

A synchronous channel provides a guaranteed quality of service (QoS) for wireless transmission, although it is important to realise that the guarantee relates to the time when a packet will be transmitted and is not a guarantee that it will be received. Synchronous channels are therefore useful where a defined or bounded latency is important. Because they are generally sent at known time boundaries, they can provide a reliable method of transmitting low latency data packets.

802.15.4 provides the same concept with its guaranteed timeslots for applications using superframes. (These are not used by ZigBee.) One consequence of guaranteed timeslots is that each slot only gives a fraction of the maximum bandwidth for the standard. In 802.15.4, each of the seven guaranteed timeslots can transmit with a bandwidth of just over 13 kbps. By comparison, for Bluetooth, each of the three SCO slots has a bandwidth of 64 kbps. Whilst Bluetooth is adequate for voice applications, in both cases, the low bandwidth means that they are only of use for data applications with low data rates.

2.3.9 Reliability

Most wireless standards use the phrase 'reliable transmission' somewhere within their text. It is applied to data channels and indicates a packet with error correction applied, associated with a protocol that will request retransmission for packets with failed CRC, so that a correct packet can eventually be delivered to a higher-level application.

For almost all applications, the word 'reliable' is a valid one, but it is important to remember that the phrase comes with a few provisos. The first is that although it is likely that the data will eventually be presented to the application layer, there is no guarantee regarding when that will happen. For any but the worst-quality links it will be close to immediate. As the link quality gets worse and more retransmissions are required, it may incur a significant delay. If your application needs data to be delivered on time, then it may be necessary to consider a number of different strategies. One may be to use a synchronous channel and accept that some packets may be lost. Using a better radio could improve the link budget and result in more reliable transmission. Or if latency is less important than relative arrival time, the data can be time-stamped and reordered into a correct temporal order at the receiver. It all depends on the application.

Taking a semantic view, the word 'reliable' in a wireless context is not necessarily what might be expected. A reliable link is not absolutely reliable, but depends to a large degree on the length of the CRC used to protect the packet contents. An n-bit CRC does not provide total protection – it provides a check that the packet is not corrupted, but does not protect against that 1 in $2n$ case where the corrupted CRC happens to match the calculated one. For a 32-bit CRC, that means that at a continuous data rate of 1 Mbps, you can expect a bad packet on average once every hour. Developers of applications that transfer files of this size have come across this problem and puzzled over the appearance of errors in what was described as a reliable connection. If you do need to transmit very large files over wireless and require total reliability, then you should consider standards that use long CRCs, or add application-level error control.

2.3.10 Audio and video

The transmission of audio and video has to be considered separately from voice. They require significantly higher data rates than can be achieved with uncompressed transmission of the stream (at least with the wireless technologies we are considering within this book), therefore they resort to sending compressed data. The compression involves encoding and decoding the data at each end of the link, which adds delays. As a consequence, there is no advantage in using a synchronous channel, so encoded audio and video data is sent over an asynchronous link. As there are a multitude of different, incompatible codecs used within the market, profiles are defined by the standards bodies, which define the encoding types that are supported and how the two ends of the link will negotiate them. The aim is to provide at least a base level of interoperability for the customer. In Bluetooth, these can be found in the advanced audio distribution profile (A2DP), and in Wi-Fi, within the wireless multimedia (WMM) extension.

The most common encodings use the various MPEG standards – typically MPEG2 layer 3 (MP3) for audio and MPEG2 and MPEG4 for video. All of these increase the processing required at both ends of the wireless link and also impose a requirement to buffer several seconds of content. As a result, power consumption is significantly higher than it is for voice applications. The higher data rates also result in a lower range (see above) than for the equivalent voice link. When using MPEG codecs, there is a license fee that implementers will need to pay. This may be covered by the chipset or firmware being used, but it is best to check.

Designers using wireless for audio transmission to complement a video signal should be aware of the problems of lip-synch. The delay resulting from an encoded wireless audio link is generally long enough to be noticeable, so a user watching video on a PC screen and using a Bluetooth or a Wi-Fi headset to listen to the accompanying soundtrack will see a perceptible delay. The usual solution is to attempt to characterise the delay and then use this to offset the presentation of video on the display.

2.3.11 Usability and commissioning

It is important not to forget usability and commissioning as a part of the selection of a wireless standard. By the time you've finished your product design you'll have learnt a lot about how wireless works. Don't forget that your customer has not had this opportunity. Their usage and interoperability knowledge will be where yours was before you embarked on designing your product. If they are going to find it useable, you'll need to distil your knowledge into the user interface, preferably hiding all of the complexity.

A few lucky manufacturers, who are in the position where they control all parts of their wireless ecosystem, can set up matched pairs of products in the factory and deliver a system that is ready to go. Most wireless products, however, are designed to be interoperable with a potentially unknown number of products from different manufacturers. They need to be installed, sometimes by a trained engineer, but more frequently by a non-technical consumer. The choice of technology, the number of profiles or applications supported and the user interface design have a very great bearing on how successful that will be. It is an area that must not be ignored. A number of manufacturers of wireless products have seen customer return rates in the range of 30–40% of product shipments because they've only considered this at a late stage of design.

As a general rule, the simpler the topology, the easier it is to set up a wireless product. The classic example of ease of use is a simple wireless keyboard, where both ends of the link are supplied in the box – the keyboard and a USB dongle. They are connected by simultaneously pressing a button on each. Such simplicity is generally only available where one manufacturer supplies both ends as a matched pair. For other applications, more complex user interfaces are required.

Where a standard includes a connection scenario, as with Wi-Fi's protected set-up, or certain of the ZigBee and Bluetooth profiles, it is important to adhere to it, rather than attempting to redesign it. Users become used to a base level of market interfaces and it is often better to support these rather than push a new, unknown one at them. It also important to use the features of the latest release of the standard. Most standards bodies have come to learn the importance of interoperability and usability and with each new release have included more tools aimed at simplifying this for the user.

Finally, don't forget that things will go wrong. For many applications, customers, will change one of the products that form part of their wireless infrastructure as it dies or a more desirable one comes onto the market. When that happens it's important that they can reconfigure the remaining products and connect the new one to it, preferably without reference to the original manual, which will have been lost several months before.

2.4 Conclusion

This chapter has given you a grounding in the parameters of wireless standards. Its purpose was to allow you to ask informed questions to help you decide which standard to use. There is frequently no obvious choice, and often the best choice may not be the first option that comes to mind.

All of wireless is a compromise. The final choice may be based on technology; it may equally be based on another product that you wish to connect to. What is important is to understand how the different aspects of performance relate to each other, so that you can find the compromise of best performance and lowest cost.

The next chapter looks in more detail at security. It digs a bit deeper for those who want to understand what the threats and implications of security really are. Once that is done I'll look at

each of the different wireless families in turn, before explaining
how to get the best performance out of your chosen standard.

2.5 References

[1] Nick Hunn, What's involved in providing a 1 km Bluetooth link?
New Electronics, (September 2007), www.newelectronics.co.uk/
article/11504/Stretching-Bluetooth.aspx.

[2] Darren Murph, $318 Wi-Fi network bridge connects two locations
up to 5 miles apart. *Engadget*, (May 2008), www.engadget.
com/2008/05/22/318-wifi-network-bridge-connects-two-locations-
up-to-5-miles-ap/.

[3] ZigBit amp module. Meshnetics, (2009), www.meshnetics.com/
zigbee-modules/amp/.

[4] The ANT Alliance, www.thisisant.com/.

[5] The Z-Wave Alliance, www.z-wavealliance.org/modules/
AllianceStart/.

[6] The EnOcean Alliance, www.enocean-alliance.org/.

3   Wireless security

Wireless security may come as something of a shock to designers used to wires, where security is always assumed to be implicit. There, you connect a cable between two sockets, data are transferred and the assumption is that no one can or will intercept the signal. The wireless world is very different. Life is far more complex than it is within wired systems. Not only is there a question of ensuring that the correct devices (and only those correct devices) are connected to each other, but it is also much easier to intercept the wireless data stream, so that also needs to be secure.

Every wireless standard has started off with this knowledge and has produced specifications that attempt to provide security similar to that experienced with a wired system. Most failed to provide an adequate level of security, at least in their initial attempts. To a large degree, the blame for this can be levelled at governments, particularly the US Government. Until recently, there has been a level of paranoia about exporting any encryption technologies that made it difficult for security agencies to intercept and crack messages. As a result, standards bodies, particularly those based in the USA saw little point in writing security specifications that would have made the products that implemented them illegal to export. In more recent times, there has been a relaxation in export controls, allowing standards to embrace higher levels of security. Nevertheless, prospective manufacturers of wireless equipment may still find that they need to apply for export licences for what they consider to be everyday, consumer devices (see Chapter 10 for more details of export controls).

As a result of these pressures, early implementations of wireless security were far from perfect and presented a tempting challenge for those wanting to break them. Cracking any new wireless security protocol has been a fertile ground for hackers and academics, who see it as a way of increasing their prestige. Unfortunately, the possibility of capturing confidential information from a wireless transmission, however academic that possibility may be, is an appealing subject for journalists, who have publicised the flaws in each standard as they have been uncovered. Wi-Fi has been a particularly successful target for them, not only because of some weak initial security schemes, but also because of the availability of hacking tools within the Windows and Linux communities. As we shall see later, the availability of hacking tools is an important step in the evolution of good security within a standard. Their existence does not mean that a standard is insecure, in fact it often means the opposite, but it is an important step on the road towards security.

As a result of adverse publicity, the perception of wireless networks as being potentially insecure is widespread. Bluetooth and Wi-Fi have done much to upgrade the level of security they provide. These enhancements have been subjected to extensive scrutiny and attack. At the time of writing, no serious flaws have been uncovered with either, at least against compliant implementations of the latest versions of each standard.

ZigBee Pro and Bluetooth low energy are less well tested, simply as a consequence of their lower market exposure. Both currently have limited deployments, much of which is within embedded applications. As a result, they have not yet become a serious target of the hacking community (in its positive sense). All of these standards take security seriously and there is every indication that they have learnt from the shortcomings of the past; however, the ultimate proof of their security will come only when they are deployed in hundreds of millions.

There is certainly no intention to suggest that the writers of wireless specifications are less than diligent about security, but security is a chicken and egg game played between the standards bodies and those who want to prove that the standards are inadequate. It is likely that even the best security standards we have today will be cracked at some point. That does not invalidate the standard – those who use wireless just need to be aware of the issues and make their design and deployment decisions accordingly.

3.1 Security attacks

Wireless security needs to be considered at a number of potential levels, each of which provides the opportunity for an attack.

3.1.1 Discovery

At the most basic security level, wireless devices can control whether or not they can be 'discovered' by other devices, which may include wireless monitors or scanners. The less time during which a device makes its presence known, the less likely it is that it will be visible to an attacking device. This can be particularly important in the initial connection stage, when device-to-device security is being set up. During this phase, messages are likely to be sent unencrypted between devices whilst they negotiate their security keys. This small space of time, normally only a few hundreds of milliseconds, which occurs only a few times during the product's life, is when wireless connections are at their most vulnerable.

3.1.2 Eavesdropping (interception)

Once devices are operating, techniques such as frequency hopping, power control and widely spread signals can be employed, which make it more difficult to intercept packets that are being transmitted. Although these cannot be changed within a standard, they may affect the choice of standard for a specific task. Obviously, the more difficult it is to capture a packet of data, the more secure a transmission is likely to be.

3.1.3 Denial of service

Denial of service (DoS) is an attempt to prevent a radio from receiving an intended message. Jamming is the simplest DoS attack and consists of a high-power signal that overloads the receivers of the radios being targeted, preventing them from receiving any other transmissions. All radios can be jammed, but some are more resilient than others. Although denial-of-service attacks can be used deliberately to block wireless networks, they can also happen without malicious intent when several uncoordinated devices are transmitting in close proximity using the same spectrum.

Denial of service also covers more subtle attacks, where an attacker bombards a wireless network with manipulated versions of packets, which are processed by the receiving device. This attack is more subtle than jamming. The intent can be twofold. The simpler attack is to overload the device with data that it believes are valid, which results in it exceeding its processing capacity, which may cause it to stop. A more subtle attack is to send deliberately malformed packets in the hope that a device malfunctions when it tries to interpret these. If this happens, then the attacker may attempt to use this malfunction as a way of gaining access to the device, allowing it to penetrate further up the stack. This attack can be effective where there is a known flaw in a wireless stack implementation, which allows a malformed packet to crash or circumvent a security feature of the device. This particular form of attack has been publicly observed with some early Bluetooth, 802.11 and 802.15.4 implementations. It is important to note that this form of attack is not due to any shortcoming in the standard itself, but to an implementation of the standard which does not correctly handle erroneous packet formats. Developers should make sure they determine the robustness of stacks that they choose to use in their designs.

3.1.4 Man-in-the-middle attacks, spoofing and bluejacking

A major source of interest to researchers is the subject of man-in-the-middle attacks, where a wireless device masquerades as a trusted device, normally during the initial connection phase, to 'take over' the connection. To perform this attack, a device 'pretends' to be a valid unit for a wireless device to connect to, masquerading as the product that a user is trying to find. Typically, the man-in-the-middle device will carry out a connection procedure with the initiating device, followed by it passing on its own credentials to the target unit. The first device thinks it is connected to its target device, but instead the connection is a two-part one, with all traffic going through the surrogate unit in the middle.

Once a connection is made, all of the transmitted data from each device pass through the man-in-the-middle node, which has the keys to decrypt the data and siphon them off. If the attacking device is intelligent, it can retransmit the signal to the target device, so that neither end of the link are aware that someone is listening in to their conversation. The only way that the transmitting device may realise that it has been compromised is if the man-in-the-middle node is missing. In this case, the initiator will discover that it has no connection with its target device when it tries to contact it, as all transmissions and security connections are with the missing man-in-the-middle node. By that time, important information may have been stolen.

A related security issue is spoofing, where a device pretends to be a valid connection. This is most common in Wi-Fi networks, where a device emulating an access point 'pretends' to be a valid access point by broadcasting a common SSID (service set identifier – the name which appears when you scan for access points). This is most commonly attempted where connections are made without security, as is usual with public hotspots. The access-point software will be written to look like that of a valid hotspot and will request that a user logs in by providing their credit card details. At this point, it will typically tell the user that their card has been refused and to try another card. More sophisticated frauds may connect the user to a valid Internet access point using their credentials, so as not to arouse suspicion. In either case, the fraudulent hotspot will appropriate the user's credit card information and sell it on.

Bluetooth has received attention from a similar unwanted connection in an attack known as bluejacking. Like Wi-Fi hotspot spoofing, it is not a product of inadequate security within the standard per se, but of the relaxed security associated with making an application easy to use. Once again, it plays on the lack of understanding that a user has of the implications of an open wireless link.

Bluetooth allows devices to be permanently discoverable, and for content to be transferred between devices on an ad hoc peer-to-peer basis. Many mobile phone users consider this to be one of the most useful features of Bluetooth, using it to share photos, music and ringtones.

On smartphones, which run applications, the same process can also be used to transfer these applications. A number of applications that contain viruses have emerged, which use this ability by searching for other phones advertising a Bluetooth connection and sending copies of themselves to these phones. Once installed, they can run malicious programs, allow unauthorised access to content on the phone, or make calls to premium numbers.

Phone manufacturers try to counter these by preventing automatic installation of applications received in this way and requiring that users confirm that they want to install them. Research shows that a significant percentage of users continue to install these viruses. In one instance, a virus was written that displayed the message, 'Do you want to install this virus?' Around 30% of users decided that they did. Once installed, the virus will begin to scan for other new phones to infect.

These two examples illustrate an important dilemma. Wireless can open a device to vulnerabilities that are new and different from those experienced on wired devices. Wireless security by itself may not necessarily prevent these. To safeguard the user it needs to be complemented with well designed user interfaces and education.

In both of these cases it can be argued that this is not a flaw in wireless security, but just a new type of phishing scam, which is enabled through a user's lack of understanding of wireless. Both are easiest to perform where there is no security enabled on the wireless connection. It illustrates the eternal dilemma of security – adding security makes a system more difficult to use, so in applications like public hotspots, it is turned off. If it were enabled, most users would need to sign up for security certificates before they travelled, which would result in an unacceptable lack of income for the operators of hotspots. That would probably make them uneconomical, so that they would disappear.

3.1.5 Address tracking

Although not a security flaw that compromises the data being transmitted over the network, address tracking is a concern where there is a desire to conceal the location of a device (or the person carrying it). It arises because digital devices normally transmit their address in the clear (unencrypted), so that an overheard transmission will yield up the uniquely identifiable address for that device. For short-range wireless devices, this will only be detectable within a few tens or hundreds of metres of a device. Nevertheless, in theory this allows a mobile device to be tracked if a range of sensors are located along its likely path to scan for its presence. Although this is at the more paranoid end of security concerns, there are some users who consider this to be an unacceptable invasion of privacy. Some standards attempt to prevent address tracking by allowing private, rotating or anonymous addresses.

3.2 Security features

To address these attacks, wireless networks apply a number of standardised processes to provide an adequate level of security. These are shown in Fig. 3.1.

3.2.1 Authorisation

Authorisation is the process of finding other devices that want to make a connection. It starts with discovery, where a device announces its presence directly, or responds to a device scanning for discoverable devices. Limiting the time that a device is in this state is an important security step, as ideally a device should only be able to advertise its presence for a limited time, when it knows that it is within the vicinity of a valid device wanting to connect to it. However, as we saw above, in some cases a Wi-Fi hotspot or a mobile phone may want permanently to advertise its presence, to enable particular usage models.

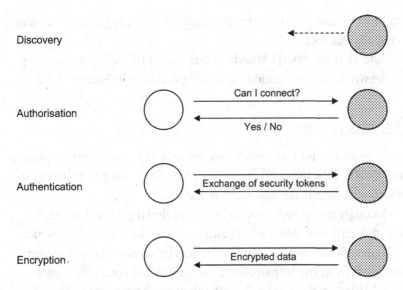

Discovery

Authorisation

Can I connect?

Yes / No

Authentication

Exchange of security tokens

Encryption.

Encrypted data

Figure 3.1 The security process

In the wired world analogy, authorisation can be viewed as the equivalent of observing that a device has a plug or a socket and thinking that you might have a suitable cable to connect it.

3.2.2 Authentication

Once two devices are authorised to talk by discovering each other, they should initiate an authentication procedure. This allows them to prove their identity to each other using a form of challenge. It is a crucial step, where devices should be able to confirm that they have the right to connect to each other and the ability to demonstrate this to the user. It is the starting point for a negotiation of security features. Typically, this will include setting up a secure (encrypted) link and using it to exchange encryption keys for use in future communications, which may be used to enable higher levels of security. At the end of the authentication process, devices should be ready to provide a secure communications link.

Authentication is a vitally important part of wireless security, as it ensures that the correct devices are connected to each other. It can be considered as the wireless equivalent of plugging in a cable to the correct socket.

(Note that the 802.11 standard has a slightly different meaning for the word 'authentication', which is explained in Section 5.2.2.)

3.2.3 Encryption

Authorisation and authentication provide the equivalent of plugging a cable into the correct socket. They do nothing to protect the subsequent data transmission over the link.

Although it is possible to detect data flowing over a cable, it is very difficult and generally requires access to a building, which itself provides a layer of security. Hence, in normal network design, data theft from the transmission medium is not normally seen as a threat. Wireless data are different – they can be captured. This may be easy or difficult depending on the characteristics of the radio, but it is always possible and in many cases can be done from outside the building where the wireless devices are being used. Moreover, it can often be done without specialised equipment. Hence, in the wireless environment it is important to protect your data against eavesdropping by encrypting them.

All of the standards address this issue by providing methods for encrypting the data being transmitted. During the initial authentication process, link keys are exchanged, which are used to encrypt the contents of subsequent data packets. At the receiving end, these keys are used to decrypt the information.

It should be noted that the encryption embodied in the wireless standards only covers the wireless link. After decryption at the receiving end, the data are no longer protected. This is often forgotten, and has resulted in a number of cases where data are intercepted where they are unencrypted on the backbone network. If an application requires end-to-end security, then implementers must realise that the wireless link is not the only point where data may be stolen. In these cases, additional protection needs to be provided at the application level.

Where data are sent wirelessly over multiple hops across a network, care needs to be taken to ensure that end-to-end security is in place for the complete journey. Implementing security on a per-hop basis can be very intensive in terms of processor power and can open up security loopholes.

Encryption techniques within wireless standards normally only cover the data packet payload. Headers may be unencrypted. Developers should not try to use spare fields outside the data payload for sending sensitive information, as they are potentially visible to anyone capturing the wireless traffic.

3.2.4 Other features

Although standards try to ensure that the initial authorisation and authentication process is secure, they always require some degree of unencrypted information to be sent over the air, which could be intercepted. If this is a concern, there are two basic options. The first is to ensure that no one could intercept the initial connection process. That can be done by ensuring that it takes place in a location that is not accessible, or which is screened to prevent interception of the signals. An alternative, more practical, approach is to use a non-wireless method to perform this initial connection.

For this reason, most security schemes allow the option for these initial phases to be performed by some means other than the wireless link. This is referred to as 'out-of-band' or OOB.

Common techniques for out-of-band authentication include:

- Preprogramming security information at manufacture and shipping devices preconfigured to work with each other,
- Using a cable at commissioning time to transfer the security information,
- Using an alternative wireless technology that is more difficult to intercept. Common examples are NFC (near-field communications) or optical connections (e.g., barcode readers).

The use of out-of-band authentication gives added protection against attacks during the short but vulnerable authentication stage. It may also simplify the process for the user. However, it generally adds cost to the product and may make it less interoperable, as devices using OOB techniques are likely to be limited to connecting to other devices using the same technique. The choice of specific OOB technology is currently outside the scope of any of the wireless standards.

3.3 Generation and distribution of link keys

Complete books have been written on wireless security techniques – it is a complex subject way beyond the scope of this volume. But there are a few basics to bear in mind when implementing a design. The different standards all offer strong encryption algorithms, but these are often only as strong as the link keys that they use. To be secure, the amount of entropy or randomness in generating the key is immensely important. Where the key is generated by a user, the longer and more complex that key is, the better. Using a short or predetermined key is a sure way to degrade security. The common use of '0000' as a key for Bluetooth headsets is a prime example of how to throw away most of your protection. Unless the security procedure has an internal method of generating a sufficient level of entropy, user-determined keys should aim to be a minimum of 8, and preferably 16, alphanumeric characters.

3.4 Comparison of security procedures

It is useful to compare how each of the technologies deals with the different elements of security. Although they take different approaches, there are obvious similarities, which is not unexpected as they are trying to solve the same problems. Throughout this section we will consider the most recent versions of the standards, i.e., Bluetooth 4.0, Wi-Fi Protected set-up specification v1.0, WPA version 3.0 (including WPA2) and ZigBee PRO. Designers should always adopt the most recent security recommendations of a standard. If they differ from a previous release, it is normally for a very good reason – which is often that the previous implementation has been cracked.

Some of the approaches have evolved as a result of the different uses and topologies for the different standards. For example, Bluetooth needs to cope with multiple ad hoc connections, so its security architecture has to accommodate devices that may be battery powered, have limited processing power and little or no user interface. In contrast, Wi-Fi targets an application where more complex devices connect to an access point, which invariably has a backbone link to another network. This permits central security radius servers to issue certificates to the device, allowing a high degree of security, but requiring far more processing capability.

ZigBee combines elements of both, with a network that includes sensor nodes that have limited processing power and a central, powerful trust centre.

The complexities of combined wireless standards, such as 802.11 over Bluetooth are not covered. Designers working with these should consult the appropriate standards.

3.4.1 Susceptibility to attack

3.4.1.1 Bluetooth
Bluetooth is relatively immune to having its packets intercepted, unless an attacker is aware of its frequency-hopping sequence. That's a result of its fast frequency hopping over 79 different channels. Although the modulation of signals is relatively simple, it would require a sophisticated receiver with a wide bandwidth to cover all channels if it were to have a chance of detecting each transmission. The standard ensures that the frequency-hopping sequence cannot be determined under normal operation. Devices that are permanently left in either a discoverable or connectable mode may be more vulnerable.

Denial-of-service attacks covering the entire 2.4 GHz band will prevent Bluetooth operating. If the jamming only covers part of the band, the adaptive frequency hopping will give Bluetooth a chance of avoiding some or all of the jamming frequencies.

The most recent secure simple pairing technique, introduced in version 2.1 of the standard, provides an authentication technique that can largely remove the problem of man-in-the-middle attacks by generating a high entropy key-set and then present the user with comparison indicators to confirm that the correct devices are connected. This technique should be used for all new designs.

3.4.1.2 Wi-Fi

Wi-Fi networks and products tend to be less mobile, with fixed access points that transmit at high power on a fixed channel. They also regularly broadcast beacon signals, so intercepting signals is trivial and can be accomplished with most PC-based Wi-Fi implementations.

Denial-of-service attacks covering the channel on which a Wi-Fi network is operating on will prevent it from operating. The standard does not include any techniques to attempt to recover from jamming.

Authorisation is uncommon in Wi-Fi or 802.11 devices and is implementation-specific. In normal operation, access points use beacons to advertise their presence to all devices. Some access points allow the setting of access lists based on the MAC address of the connecting device. These can be managed from the network side or by a user who has authenticated with the access point. It should be borne in mind that devices can spoof MAC addresses, so this is not a foolproof scheme.

Wi-Fi can also suffer from the fact that the control and management frames at the MAC level are unprotected. This allows attacks, which attempt to flood the network with spurious disassociation and de-authorisation packets.

Above the MAC level, the TCP/IP stack is essentially the same as that used in wired networks, so considerable work has gone into securing most of these stacks.

Spoofing is a problem with Wi-Fi unless security is enabled. Although it might seem obvious that security should be enabled, public applications like hotspots turn it off in favour of usability. This should be avoided and the most recent WPA2 used to ensure that security is enabled. Note that although security is specified for ad hoc connections within 802.11, it is at a lower level than that which has been developed by the Wi-Fi Alliance. This should be rectified when the Wi-Fi Direct specification is completed and published.

The more recent 802.11n standard gives rather better natural protection for data packets, as its use of multiplexed signals makes them more difficult to capture and resolve.

3.4.1.3 ZigBee

ZigBee is similar to Wi-Fi, in that it transmits on a single channel. It can move channels in the presence of interference, but this is a moderately slow process, not a fast dynamic frequency hop. However, the fact that transmissions are intermittent, and power is generally lower than with Wi-Fi means that these transmissions are more difficult to detect. The latter benefit may be counterbalanced in the case of an extended mesh, which increases the area over which interception can occur.

Denial-of-service attacks covering the channel on which a ZigBee PRO network is operating will prevent it from functioning. The standard incorporates optional frequency agility that attempts to find a clean channel and move the network to that channel. Although optional, this feature should be implemented in the interest of a robust system.

ZigBee PRO has introduced protection for authorisation and authentication packets at the MAC level, closing one of the previous security loopholes. It also introduces the concept of a trust centre, which is used to authorise all new connections.

3.4.1.4 Bluetooth low energy

Like Bluetooth, Bluetooth low energy employs a frequency-hopping regime over the entire band, making it difficult to detect more than a few packets. Although the hopping sequence is slower, it benefits from the intermittency of transmission that comes from low-power duty cycles, as does ZigBee. Taken together, these make Bluetooth low energy secure against interception.

Bluetooth low energy implements a whitelist function at a base-band level that can be set to reject connection requests from any address that is not known. Although its primary function is to prevent the host controller from being woken by irrelevant incoming packets, it can also be used to provide protection from unauthor-ised devices.

Bluetooth low energy does allow broadcast packets to be trans-mitted on three fixed advertising channels that contain data with no encryption. These are intended for public messages, such as time, temperature or local news. They should never be used for sensitive data.

Denial-of-service attacks covering the entire 2.4 GHz band will prevent Bluetooth low energy from operating. If the jamming only covers part of the band, the adaptive frequency hopping will give Bluetooth low energy a chance of avoiding some or all of the jam-ming frequencies. Bluetooth low energy is susceptible to a coor-dinated denial-of-service attack on its three advertising channels. However, as these are spread across the entire spectrum, any such attack is likely to use a full band jammer, which would disrupt any of the other 2.4 GHz wireless standards.

Bluetooth low energy security has evolved from that of Bluetooth, which is well tested. Despite this, it is a new standard and early implementations may well contain loopholes. The first release does not include secure simple pairing, so care should be taken to avoid man-in-the-middle attacks during initial connections.

3.4.1.5 General

The previous sections provide a statement of what is contained in the relevant standard. Many security failures come not from the standard, but from an incomplete or erroneous implementation. Often that occurs because the behaviour, when invalid packets are received, is not sufficiently defined. Just because a standard has fixed a problem does not mean that an implementation has reflected that change correctly, or that an implementation has not introduced its own security holes. A common source of problems is with stacks that enable attacks that exploit buffer overflows or underflows. In the light of this, try to ensure that the stack you choose is well tested and keep up to date with any security information relating to it. As a general rule, stacks and standards that have been in the market for a longer period of time are likely to be more robust.

More determined hackers can spoof the unique MAC addresses of wireless devices to mimic other devices in the area. As long as security is employed this should not be a problem, as access will be denied unless the device has a correct link-key.

All wireless standards, with the exception of Bluetooth low energy, which has introduced the concept of random addresses, are susceptible to address tracking. However, it is unlikely that this is more than an academic security concern for any commercial application.

3.4.2 Security implementations

3.4.2.1 Bluetooth

Bluetooth devices make their presence known to each other by a process called discovery. To make a connection, the master device enters an inquiry scan mode, during which it broadcasts inquiry messages on its inquiry scan channels.

Depending on their application, devices can constantly advertise their presence and capabilities, or remain non-discoverable (hidden) until placed in discoverable mode, which normally lasts for a few tens of seconds. During this period, when a connectable device is actively listening, it can respond to an incoming inquiry packet.

Once a device is discovered, the master can request a connection by using a page-scan channel. This is directed at the device discovered during the inquiry scan and starts a procedure where further information is exchanged between devices, which will determine whether to set up a connection.

The discovery process at both ends is normally entered manually, either by a user action on a device with a user interface, pressing a button, or simply applying power to the device for the first time. How this is done is implementation-specific. Some applications will allow devices to be set as 'always discoverable'. This is commonly found on personal devices, where users may want to opt in to ad hoc services. Most are unaware that this compromises security.

The secure simple pairing process of version 2.1 and above uses elliptic Diffe–Hellmann public key cryptography to generate link keys on two devices with an equivalent of 95.3 bits of entropy. (This is chosen to be slightly better than the maximum 16-bit alphanumeric PIN used in previous versions of the specification.) Unlike earlier versions, where the strength of the encryption depended on the randomness and length of PIN entered by the user, the full entropy is achieved every time using this scheme.

Man-in-the-middle attacks are prevented by providing feedback mechanisms to the user to allow them to check that the correct two devices have been paired. In the case of devices with displays, this takes the form of a random six-bit number, which is indicated on both devices. If the number is the same, there is a 1 in 10^6 chance of a MITM attack having occurred, which is more than adequate. For devices without a numeric display, alternative comparison messages can be used, such as flashing lights. The fewer states presented to the user, the smaller the differentiation provided against the possibility of a MITM attack. The standard also supports a number of alternative confirmatory techniques, including 'just works', out-of-band and passkey entry.

3.4.2.2 Wi-Fi

Wi-Fi access points, or masters in an ad hoc network, advertise their presence by sending out regular beacons containing their SSID. There is generally no attempt to conceal their presence, nor any mechanism to prevent devices attempting to connect to them.

When a device discovers a Wi-Fi access point or ad hoc master, it starts its authentication procedure. In early versions of Wi-Fi, this used wired-equivalent privacy (WEP). Wired-equivalent privacy authentication uses a clear text message from the client, which is encrypted and returned using a pre-shared key. However, it is relatively easy to crack this by listening to the returned authentication frames. Wired-equivalent privacy has been replaced with authentication processes based on the IEEE 802.11i recommendations, using either pre-shared keys for general purpose (personal) use cases, or stronger port-based authentication, requiring a back-end authentication server for more demanding (enterprise) applications. Enterprise authentication uses the extensible authentication protocol (EAP-TLS) and is widely used in corporate environments. This does impose significant processing and memory requirements on client devices and may not be compatible with low-power, portable applications. It is generally agreed that these schemes provide sufficiently strong authentication.

The first approach to encryption also used WEP, employing 40- or 104-bit pre-shared keys, which are concatenated with a 24-bit initialisation vector to generate a 64- or 128-bit key. The short length of the initialisation vector made this vulnerable to attack, with numerous hacking tools rapidly appearing on the web.

The Wi-Fi Alliance replaced WEP with Wi-Fi Protected Access (WPA), utilising a portion of the 802.11i security standard and increasing the key length to 128 bits. This in turn has been cracked, leading to today's security solution of WPA2, which comes in personal and enterprise variants. It is based on the robust security network of IEEE 802.11i, which uses 128-bit AES encryption. To date, this has not been broken.

3.4.2.3 ZigBee authorisation

Prior to the release of the ZigBee PRO specification, security in ZigBee was weak. The PRO release includes both standard and high-security modes, of which the latter should be used unless there is a good reason for implementing the weaker standard-security mode. The high-security mode gives ZigBee PRO one of the most secure schemes of any of the wireless standards.

Authentication is performed using a pre-shared key that is fed into an elliptic curve Menezes–Qu–Vanstone key-establishment mechanism (ECMQV). All that needs to be said is that this is an efficient and secure means of authentication. As a further refinement, some ZigBee application profiles, notably the smart energy profile, mandate additional security features, including the use of the Matyas–Meyer–Oseas hash function to generate the preconfigured key.

Despite the strength of this arrangement, the level of security is badly diminished when designers perpetuate the practice of using trivial PIN codes for customer installations.

Encryption of the data takes advantage of the 128-bit advanced encryption standard (AES) defined within the 802.15.4 MAC that sits underneath the ZigBee stack. This is widely considered to be the best general-purpose encryption technique available today.

ZigBee PRO also includes the useful feature of secure tunnelling encryption. This allows data to remain encrypted as it passes through any number of nodes within the mesh network before being decrypted by the destination node. Without this feature, security in a mesh network would require far more processing capability within each node.

3.4.2.4 Bluetooth low energy

Bluetooth low energy has defined a security-management system specifically designed to try and meet the needs of a standard where the peripheral devices have less processing capability than the collector device. It is optimised so that the memory requirements of a responding device are lower than those for an initiating device.

Security management is a three-phase process, starting with pairing, followed by short-term key generation (STK) and completed with distribution of a transport-specific key.

Three options are available for the pairing process, where devices exchange information, including how to proceed in the second stage. Three alternatives can be used – 'just works', passkey entry and out-of-band. Distribution of a transport-specific key is optional, but if used, must be performed over an encrypted link.

'Just works' should only be used on the simplest device, with little or no security requirement. It offers no protection during the pairing process against man-in-the-middle or eavesdropping. Although these are unlikely, and the subsequent encrypted links may be secure, users will not be able to tell if the pairing process was compromised.

Passkey entry requires a six-digit passkey to be entered to confirm the pairing process. Static passkeys are not allowed. If a device has no user input capability, but contains a display, then it must present a random six-digit number, which the user enters on the other device.

Bluetooth low energy differs from Bluetooth in using 128-bit AES CCM to encrypt payload data in a secure link.

3.5 Testing security – in praise of hacking tools

A few years after products are deployed that use a wireless standard, there is an inevitable announcement of the release of a set of hacking tools. This is generally greeted with one of two emotions – either fear and loathing, or relief. I prefer the latter reaction, as a set of tools and even hacks from an external group is one of the best ways of testing implementations and providing a means of putting right what is wrong.

The groups publishing these tools range from security experts, through stack developers themselves, to academics. They approach the standard with a different viewpoint from the original specification writers and often manage to find loopholes that were missed. After one or two iterations of specification release and hacks, the security generally tightens up to the point where it is more than adequate for commercial applications.

Hacking tools are also useful for designers as they can be used for testing the robustness of implementations from different suppliers.

Most of the Wi-Fi hacking tools are available at the Wi-Foo portal,[1] which also contains a comprehensive range of Bluetooth hacking tools.[2] The authors of this site have also produced a good book on the subject of wireless vulnerabilities.[3] ZigBee hacking tools have only just started to appear, but interest in developing them is growing.[4]

3.6 References

[1] Wi-Foo, Recon and attack tools. www.wi-foo.com/index-3.html.
[2] Wi-Foo, Bluetooth security tools. www.wi-foo.com/
    ViewPagea038.html?siteNodeId=56&languageId=1&contentId=-1
[3] Andrew Vladimirov, Konstantin V. Gavrilenko and Andrei A.
    Mikhailovsky, *Wi-Foo: The Secrets of Wireless Hacking* (Pearson,
    2004).
[4] Joshua Wright, *KillerBee: Practical ZigBee Exploitation
    Framework* (2009) www.willhackforsushi.com/presentations/
    toorcon11-wright.pdf.

4   Bluetooth

Bluetooth [1] started life in 1998, when it was announced by an industry consortium of five companies – Ericsson, IBM, Intel, Nokia and Toshiba. It was based on an earlier Ericsson development known as MC-link and developed as a wireless technology whose primary purpose was to bridge the worlds of the phone and PC, offering a connection that was low power and which could handle voice and data. The aspirations were to find a compromise of low implementation cost, resistance to interference, ease of use, interoperability, low power, voice support and good data transfer rate. The data transfer rate was set at 1 Mbps, which was significantly higher than the speed of available wired and wide area links at that time.

The standard has moved through a number of versions since its original release, as outlined in Table 4.1.

The number of releases is typical of any wireless standard, with growing maturity and stability as the specification matures. At the time of writing, versions before 1.2 have been deprecated and any new products must be based on version 1.2 or above. Having said which, version 2.1 introduced a major advance in security, which not only minimises the possibility of man-in-the-middle attacks, but also provides a toolkit for simplifying the initial pairing process. Any new design should consider this as the base version to implement.

4.1 Background

Bluetooth was developed from earlier proprietary radios to provide a low power, short-range radio link, whose primary application was for use with mobile phones. It emerged at the stage where

Table 4.1 *History of Bluetooth releases*

Version	Date	Key features
1.0	5 July 1999	Draft version
1.0a	23 July 1999	First published version
1.0b	December 1999	Bug fixes
1.0b +CE	November 2000	Critical errata added
1.1	February 2001	The first solid release and the basis of growth. This version was ratified by the IEEE as the 802.15.1–2002 standard.
1.2	November 2003	Included adaptive frequency hopping. Added eSCO for better voice performance. Ratified by the IEEE as 802.15.1–2005, although no subsequent releases have been added to it, leaving the 802.15.1 standard as an orphaned specification.
2.0 + EDR	November 2004	Added enhanced data transfer rate to increase throughput to 3.0 Mbps
2.1 + EDR	July 2007	Added secure simple pairing to improve security and usability
3.0 + HS	April 2009	Added 802.11 as a high-speed channel, boosting rates to 10 Mbps and above
4.0 + HS	December 2009	Includes Bluetooth low energy

mobile phones were making two important transitions: they were beginning to support data applications, which required a connection to a PC, and they were becoming a high-volume consumer item. This led to the key requirements for Bluetooth 1.0:

- Relatively low power consumption,
- Support for voice transmission to headsets,
- Support for data connections to PCs,
- Low-cost, single chip solutions.

The specification development concentrated on developing a robust, efficient and, most importantly, cost-effective standard. At a time when there were no single-chip CMOS radios, and a fair proportion of the silicon industry thought that they were not physically possible, the specification tried to define technology that could be mass produced within a few years.

Implementation cost was a major factor for the voice portion of the specification, which rejected the use of IP-based voice because of the power and cost overheads that they would impose on a headset. Instead the decision was made to transmit voice in real time, with simple digitisation.

Through the participation of Ericsson and Nokia, Bluetooth benefited from a high degree of RF expertise, resulting in one of the most robust wireless standards at 2.4 GHz. It is the only 2.4 GHz standard that has defined its own radio.

One of the key achievements of the Bluetooth SIG was a concerted lobbying effort to persuade regulatory bodies around the world to bring their national requirements for the 2.4 GHz ISM band into alignment with each other. Although the 2.4 GHz band was available throughout the world, the available spectrum, allowable modulation techniques and power limits varied widely between countries, making it impossible to ship a single product globally. The effort of the SIG's regulatory group in driving standardisation has brought almost all countries into alignment (France remaining the notable exception, where transmit power is limited to 10 mW for outside use), benefiting all standards that use this band.

Unlike other wireless standards, which build on external specifications, the Bluetooth SIG has produced a standard encompassing every layer from radio to application profiles (see Fig. 4.1).

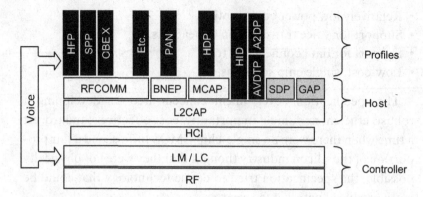

Figure 4.1 The Bluetooth stack (see later sections in this chapter for the acronyms)

4.2 The radio

When Bluetooth came into existence, it was already clear that the 2.4 GHz band was going to be used for a wide range of applications. To ensure that Bluetooth would be as robust as possible in a spectrum that was likely to become increasingly busy, the decision was made to use frequency hopping. This is particularly important for a standard supporting real-time voice, where there is little opportunity for retransmission of packets.

Within the ISM band of 2.4000–2.4835 GHz, Bluetooth divides the spectrum into 79 evenly spaced 1 MHz channels, starting at 2.402 GHz and finishing at 2.480 GHz. (There is a lower guard band of 2 MHz and an upper guard band of 3.5 MHz – Fig. 4.2.) The radios hop amongst these 79 channels at a rate of 1600 hops per second in a pre-agreed pseudo-random sequence, which is derived from the master's Bluetooth address. The symbol transfer rate of the radio is 1 Mbps, and in its first incarnation, this provided a maximum data throughput of 723 kbps.

Hopping leads to the use of timeslots to control communications between devices. Given the fast hopping rate of 1600 hops per second, or just 625 μs per slot, this imposes tight timing requirements to ensure that all devices are synchronised with each other.

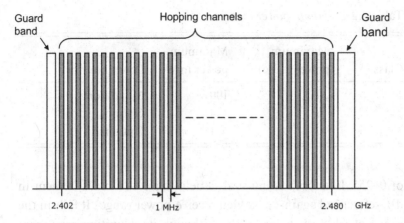

Figure 4.2 Bluetooth spectrum usage

The basic version of the standard (in version 1.x specifications) uses a Gaussian frequency-shift keying (GFSK) modulation scheme with over-the-air signalling of one million symbols per second, equating to an actual over-the-air data transfer rate of 723 kbps. Although this may seem slow in relation to today's data transfer rates, at the time Bluetooth was designed, the maximum data rate available for most users on a GSM network was 14.4 kbps for high-speed circuit-switched data (HSCSD).

As a number of different applications were envisaged, with different ranges and power requirements, the standard defines three different power classes for Bluetooth radios. These are shown in Table 4.2.

Power control is mandatory for Class 1 devices and is recommended for other classes to conserve battery life. Once two devices are connected, they will negotiate the lowest appropriate power for the connection based on a received signal strength indication (RSSI). If the measured RSSI figure falls during a connection, the device can request that the power level be increased.

The RSSI is commonly used as a measure of the strength and, by implication, quality of a received signal. In many cases, the RSSI has no units and no formal definition. Instead, it is a value that is vendor-dependent, normally being displayed as a range of 1–100

Table 4.2 *Bluetooth power classes*

Class	Maximum power (dBm)	Maximum power (mW)	Power control
1	20	100	Mandatory
2	6	4	Optional
3	0	1	Optional

or 0–255. Bluetooth is unusual in defining it as a measurement in dB, referenced against a 'golden receive-power range'. It falls in the range of −128 dB to +127 dB. Although its definition may seem arbitrary, it is normally used within a device to control its gain or operation, so has no need to be a calibrated measurement.

The Bluetooth specification defines a maximum bit error rate of 0.1%, which translates to a minimum requirement for the receive sensitivity of −70 dBm. Using this, in conjunction with the Class 1 and Class 3 transmit powers of 0 dBm and 20 dBm, results in theoretical range figures of 10 metres and 100 metres, which are often quoted for Bluetooth. In practice, the receive sensitivities of today's devices are significantly better than the base specification requirement and ranges far in excess of these are achievable. For a well designed device operating at 4 mW with an omnidirectional antenna, open-field ranges in excess of 100 metres are perfectly feasible.

In version 2.0 of the standard, a higher data rate option, called the 'enhanced data rate' (EDR), was introduced. This adds two additional phase-shift keying (PSK) modulation schemes: π/4-rotated differential-encoded quaternary phase-shift keying (π/4-DQPSK) and differential-encoded 8-ary phase-shift keying (8DPSK), which increases the symbol data transfer rates to 2 Mbps and 3 Mbps, respectively. The EDR is implemented in addition to the basic rate (BR), ensuring downwards compatibility for devices (Fig. 4.3).

To maintain backwards compatibility, only the payload section of the packet is encoded using 8DPSK. Whereas a basic-rate packet

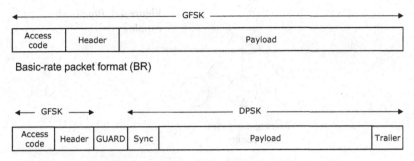

Figure 4.3 Simplified Bluetooth packet formats

only contain three main entities – the access code, header and payload, an EDR packet adds a synchronisation sequence and trailer around the payload, all of which are transmitted using 8DPSK. A guard period is also introduced between the packet segments GFSK and 8DPSK.

The Bluetooth radio does not lend itself to supporting significantly higher data rates, so in the version 3.0 release, the concept of an alternate MAC/PHY (AMP) is introduced, which allows Bluetooth to act in conjunction with another radio technology. This allows Bluetooth to perform the initial pairing and security, and then switch in the alternative radio whenever it is required for higher throughput. Version 3.0 defines a method for Bluetooth to work cooperatively with an 802.11g radio to provide ad hoc connectivity with over-the-air data transfer rates up to 25 Mbps. Silicon suppliers are now offering combination chips that integrate both Bluetooth and 802.11 for these applications.

4.3 Topologies

Bluetooth is essentially a piconet topology (Fig. 4.4), where a device acts as a master, talking to a maximum of seven active slaves. As well as simple piconets, the specification supports the concept of scatternets, where a slave device can share its time between two piconets. In theory, this allows complex star networks to be set up.

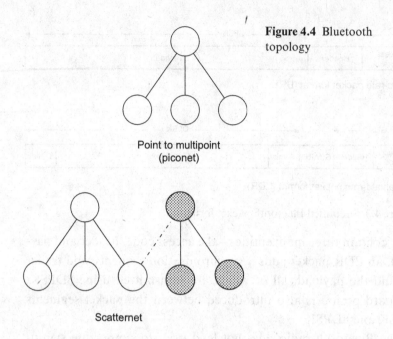

Figure 4.4 Bluetooth topology

Point to multipoint
(piconet)

Scatternet

In practice, the timing and resultant memory constraints involved in keeping track of the phases of the clocks used for synchronising frequency hopping mean that this is a theoretical rather than a practical application and scatternets have rarely appeared in commercial applications. A notable exception is the BlueCluster application from Stollmann, which uses a scatternet to build a large-scale Bluetooth mesh network.

The seven-slave limitation within Bluetooth relates to the number of active connections that can be live at any one time, which comes from the three-bit address space used for active slave addresses. Bluetooth can support larger numbers of polled connections through standard addressing, but this is rarely utilised. As with all networks, increasing the number of connected devices results in a reduction of the individual throughput, as the overall bandwidth is shared.

Much of the way in which the topology works and devices talk to each other stems from the nature of a frequency-hopping network.

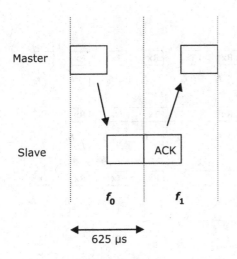

Figure 4.5 Conversing across hops

Master

Slave

ACK

f_0 f_1

625 µs

Frequency hopping naturally divides transmission up into time-slots, each of which lasts for the duration of one hop. For Bluetooth, the connection scheme starts off with 1600 equally spaced hops every second, resulting in a base timeslot of 625 µs. In the first timeslot, a master will send a message to a slave and in the following slot the slave will acknowledge the message (Fig. 4.5). In subsequent slots, the master can continue a conversation with the same slave, start talking to another slave, or go to sleep.

To allow larger, more efficient packets to be transmitted, the Bluetooth standard allows a transmission to be spread across three or five timeslots without the need to hop to the next frequency. This is illustrated in Fig. 4.6.

In effect, the frequency hops are skipped and the following timeslot is performed on the frequency that would normally have been used for that slot. This ensures that other slave devices that were unaware of the extended packet are still synchronised (Fig. 4.7).

In a crowded spectrum, particularly one where there are radios that sit at a fixed frequency (such as Wi-Fi or ZigBee), there will inevitably be channels which are more prone to interference, both to and from the Bluetooth transmissions. To improve both the performance of a Bluetooth link and help mitigate the interfering

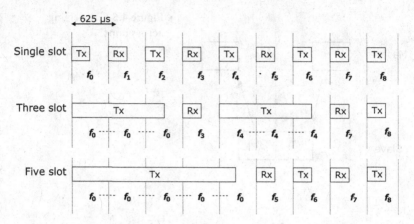

Figure 4.6 Extended hops

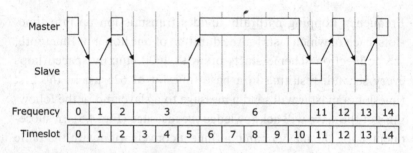

Figure 4.7 Original hopping behaviour

aspect of a Bluetooth device, an adaptive hopping scheme was introduced in version 1.2 of the standard.

Adaptive frequency hopping (AFH) works by scanning all of the channels and looking for activity on each of them. This is used to build up a picture of channels in use, which is then used to modify the hopping sequence of a Bluetooth piconet, so that these channels are avoided. This table of active frequencies can be updated dynamically to cope with variations in spectral use as well as mobility of devices. It is also possible to tell a device which channels to avoid via its host interface if that information is available at an application layer. This is most commonly the case where a product incorporates two different wireless standards.

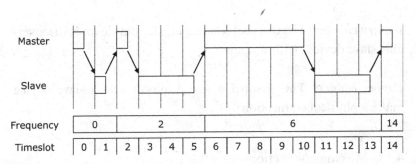

Figure 4.8 Adaptive hopping behaviour

Adaptive frequency hopping can reduce the number of used channels to as few as 20 of the original 79 channels. This can lead to a problem; if different hops are used for a message and acknowledgement, channel reuse can become too frequent. To mitigate this, the 1.2 release allows a slave to respond to the master using the same frequency as that used for the original message. Figure 4.8 shows the effect of this change.

4.4 Connections

Bluetooth is best considered as a connection-oriented system. Connections are made between a master and a slave and these connections are maintained until they are deliberately disconnected or when the link is broken, typically because a device goes out of range.

The standard describes four different connection channels, which between them cover the behaviour of a Bluetooth link.

Basic piconet channel The first of the hopping schemes described above, where the devices hop through all 79 physical channels. In most cases, this has been replaced by the adapted piconet channel.

Adapted piconet channel The scheme based on adaptive frequency hopping, where between 20 and 79 physical channels are used. Piconet channels are the only channels that can be used to transfer user data.

Inquiry channel The channel where a master device finds discoverable slave devices within range.

Paging channel The channel where a master and a slave device make a physical connection.

4.4.1 Making connections

Making connections within Bluetooth is more difficult to understand than with radios that operate at fixed frequency. The latter simply need to work through each of the static frequencies employed by the standard until a connection is made. With Bluetooth, both nodes involved in the connection are hopping around the 79 channels with their own pseudo-random hopping sequence. The hops are not even synchronised with each other. To find each other, they need to use the inquiry process.

When a Bluetooth device is first turned on it knows nothing about the hopping sequence of any other device. Each device chooses a random hopping sequence that is seeded by the value of its 48-bit Bluetooth address (BD_ADDR). (The resulting hop sequence lasts about 24 hours, at which point it is repeated.)

To find another Bluetooth device for the first time, a device needs to take on the role of a master device and use the inquiry process to look for other devices. Bluetooth devices can either be configured to respond to such an inquiry, in which case they are said to be discoverable, or they can be set so that they do not respond, when they are termed non-discoverable or hidden. It is not uncommon for devices to be set so that they will stay in a discoverable mode for a limited time. This is determined by application firmware, but helps to provide better security, as it means a device cannot enter into a transaction other than in a short, user-initiated time window. When it is in this discoverable mode, the receiving device will be executing an inquiry scan.

Inquiry mode and inquiry scans use special hopping sequences to try and ensure that there is the greatest possible chance that two devices will be able to find each other by being on the same frequency at the same time. This is necessary, as neither knows the other's hopping sequence, the frequency it is working on, or the relative clock phase of the other device. As soon as it enters the inquiry scan state, the discoverable device starts to hop at a much slower frequency – only once every 2048 hops, or once every 1.28 seconds. Independently of this, the master starts transmitting very short packets with an inquiry access code (IAC). Because these are so short, they can be transmitted twice in each timeslot, doubling the chance that they will coincide with a receive slot on the other device. These are repeatedly transmitted over 32 different frequencies, centred on a randomly chosen channel. After each sequence of 32 frequencies, the master will hop to a new centre frequency and repeat the process.

These very aggressive inquiry transmissions are efficient at finding the other device in a relatively short time. However, if any other devices in range are discoverable, it is likely that the inquiry transmission would have been received by more than just this device during the inquiry scan state. If all were to respond immediately, there is a risk that their responses would interfere with each other. To prevent this, a device that receives a successful inquiry request does not respond, but enters a random delay period before re-entering the inquiry scan mode. It now waits for another inquiry request and, upon receiving this, immediately responds with an inquiry response in the form of a frequency-hopping synchronisation (FHS) packet. The FHS packet gives the inquirer information about the other device's frequency-hopping scheme and its Bluetooth address (BD_ADDR). It also returns the device class information, which provides a top-level description of what sort of device it is, for example a phone, headset or laptop.

During the inquiry process (see Fig. 4.9), the master will discover all of the devices within range that are available for connection. It is able to determine basic information about these devices, such as the types of device they are and their friendly names. The friendly name is a text field that is normally given a default value by the

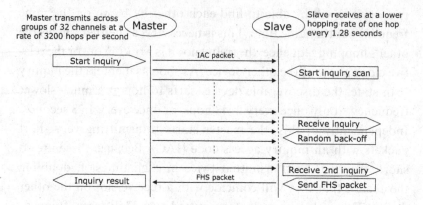

Figure 4.9 Inquiry process

manufacturer, but may be modified by the user or installer. It is useful to help identify a device before making a connection to it.

If the master (or the user) decides that it wishes to make a connection to one of these devices, it needs to use the paging channel and start the paging process.

The paging process (Fig. 4.10) is very similar to the inquiry process, but with a few key differences. The first is that it is a directed process, as the address of the target device is known. Also, because the communication is now directed at a single device, there is no risk of multiple replies bombarding the master, so the random back-off is unnecessary.

Both devices start the paging sequence using the same hop sequence as they did for the inquiry. In this case, the master broadcasts an ID packet to the required BD_ADDR. As soon as the slave successfully receives one of these packets, it responds with an ID packet containing an access code. The master then sends out its FHS packet, which contains all of the information that the slave requires to adjust its clock to the same hopping sequence as the master. It acknowledges receipt of this packet with another ID packet, exits from the page scan mode and starts hopping, synchronised to the master.

Once the master receives this acknowledgement, it starts a packet exchange that will negotiate parameters between the devices until a connection is established. This connection will then be

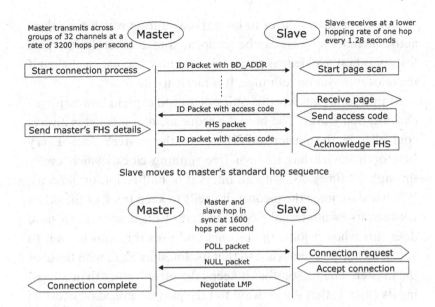

Figure 4.10 Paging process

used for subsequent piconet channel use. At this stage, the master can request more information about the services available on the device, by making a connection to the service description server on the slave. It can then use this information to make a specific connection to one or more profiles.

Inquiry and paging can take place whilst the master or slave devices are already engaged in other connections. In these cases, the time spent in inquiry or paging modes needs to be timeshared with other connections, particularly if there is an active SCO connection. Where this occurs, the time required to find and connect to a device may be significantly longer.

If a master needs to connect to more than one device, then this process must be repeated for each device. However, enough information on devices may have been captured from the inquiry scan to allow the master to move straight to the paging mode to make these connections. Up to seven active connections are permitted. This limit is determined by the active member address (AMA), which is a three-bit field used by the master to address slaves. The specification allows slaves to be 'parked', during which time their active member address can be reallocated to another device. This feature is best avoided, as parked devices may become orphaned if the master moves out of range. It is rarely used.

At this point, it is worth a quick detour to explain how hop synchronisation works. The hopping sequence of a piconet is always controlled by the master and is based on the master's clock. Every Bluetooth device has its own free-running clock, which cycles through 2^{28} timeslots, with an interval of half a slot, or 312.5 µs. When a slave joins the piconet, it needs to keep track of the offset between its own internal clock and that of the master. Each slave does this when it joins the piconet and uses this information to adjust its hopping sequence so that its timeslots align with those of the master. The reason that it keeps the offset rather than adjusting its clock is that it may want to take part in several piconets. If this is the case, it needs to keep a separate track of the offsets for each of the piconets. This can be resource intensive and may limit the number of piconets to which a device may be connected.

Because the master device transmits aggressively during the inquiry (and the paging) process, it generates a lot of traffic within the band. Hence, devices should only perform inquiries when necessary.

One final point to mention in this section is the master–slave role switch. A connection can be initiated from a Bluetooth device at either end of the link, but that device must be operating as the master to do so. In many cases, this may not correspond with which device will need to be the master in the final connection. A common example is that of the headset, where the user may initiate a connection by touching the headset. This will connect to the mobile phone, but the phone may then want to take over ownership of the link. This is allowed within the Bluetooth standard and a master–slave role is specified, which allows roles to be changed without the need to break the link. Note that some parameters, such as quality-of-service requirements, may not be retained through a role switch and will need to be negotiated once the switch is complete.

A consequence of this is that all Bluetooth devices need to be able to support all of the basic roles. This means that their capabilities and complexity are more symmetric than is the case with most other wireless standards.

4.5 Transferring data

If you read the Bluetooth standard, you'll discover that it supports a very large number of different data packet types, with names such as HVn, DMn, DVn, DMn, AUX, etc. Amongst other things, these define the size of the data payload, whether they support streaming or asynchronous data and whether they implement forward error correction (FEC) and contain a cyclic redundancy check (CRC).

Although these exist for good reasons, they are generally inaccessible to product designers. The firmware running in a Bluetooth baseband controller will negotiate these on your behalf, based on the quality of the link and the requirements of the connection.

As far as a product developer is concerned, the available data transports can be split into two different types – asynchronous and synchronous.

4.5.1 Asynchronous links (ACL)

Asynchronous link (ACL) connections are used for transporting framed data. Framed data are data submitted by an application to a logical link control and adaptation protocol (L2CAP) channel. The channel may support either unidirectional or bidirectional data transfer. The transmitted data are presented to the application on the device at the other end of the link in the same format in which they were received from the host application. Most ACL formats incorporate FEC and header error correction (HEC) to detect and correct errors.

Asynchronous link connections provide most of the links for Bluetooth applications. Supported by profiles, they allow applications to deliver data to a Bluetooth device in the expectation that the same data will be delivered to an application at the other end of the link. Data rates of up to 723 kbps are achievable using ACL links with basic-rate Bluetooth and 2.1 Mbps using enhanced-data rate. Asynchronous links can be granted QoS by setting the associated L2CAP channel parameters.

4.5.2 Synchronous links (SCO and eSCO)

The other form of link is the SCO or synchronous channel. This is used where data need to be streamed, rather than framed. Synchronous channels can coexist with ACL channels (as a minimum, they need one to configure them), but are granted guaranteed timeslots, so that they can be sure that data will be presented at a known time and with a defined maximum latency.

A Bluetooth master can support three simultaneous SCO channels, which can be split between up to three slaves. Each channel provides a bandwidth of 64 kbps. Packets are neither acknowledged nor retransmitted. If a SCO packet is lost or corrupted it is up to the receiving application to decide what to do.

Referring back to Fig. 4.1, SCO channels effectively bypass the host stack and L2CAP. Once they are set up, data flow directly from the application profile to the baseband resource manager, minimising the latency.

Extended SCO, or eSCO, was introduced into the version 1.2 specification to make voice traffic on SCO channels more robust. Whereas SCO packets have no acknowledgement or retransmission facility, eSCO provides greater reliability by allowing a limited number of retransmissions. It works in guaranteed times slots, so there is a compromise in the maximum number of retransmissions. The underlying principle is that retransmissions are only allowable until the next guaranteed slot, at which point any packets that are still corrupt must be discarded. Packets are assumed to arrive in the order in which they are transmitted, so there is no ability to store and reassemble them. Both SCO variants are unashamedly real-time.

4.5.3 Voice codecs

Most functionality is defined within the appropriate profiles, but Bluetooth does include a number of specifications for voice codecs within the core standard. These are logarithmic PCM coding using either A-law or μ-law and continuous variable slope delta modulation (CVSD). Bluetooth chips may implement these internally as standard options, or present a PCM interface for use with an external codec.

All of these codecs are designed for 'toll quality' voice, which is the quality you would expect from a landline. They are not suitable for audio streaming, as all effectively limit the audio bandwidth to 4 kHz. Audio or stereo music applications are handled using more complex codecs, which are run over ACL links. The advantage of simpler codecs like CVSD is that they exhibit minimum latency, so there are no synchronisation issues when they are used with video phones. In contrast, music codecs can introduce noticeable delays.

4.6 The lower-layer stack (the controller)

In the preceding discussion of the radio, channels and connection topologies, I have described functionality that occurs within the lowest layers of the Bluetooth stack. Most of the connections are managed by the link manager and link controller. Together with the radio and device manager, these form the entity known as the Bluetooth controller.

Figure 4.11 illustrates the data flows through the various key components of the stack. It is an alternative way of looking at the architecture compared with the more common stack representation of Fig. 4.1, but illustrates the key data paths.

At the top of the controller is the host–controller interface. This is an interface within the Bluetooth standard that has a well defined set of API calls. Chips and subsystems that expose this interface can do so over a number of defined physical transports, including UART, USB, SD and three-wire. The USB link is by far the most common. In fact, this is the interface on Bluetooth USB dongles.

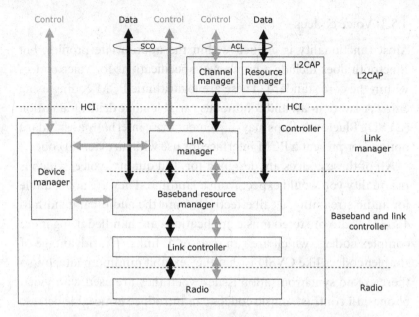

Figure 4.11 Control and data architecture

The definition of the HCI interface aims to ensure that controllers from different vendors can be interchanged. To meet Bluetooth's qualification requirements, controllers that expose an external interface at this point need to conform to the HCI specification. If a chipset incorporates higher layers, then it does not need to expose its HCI interface.

4.7 The higher-layer stack (the host)

The lower layers that form the Bluetooth controller generally take care of the connections between devices without the need for them to wake up the higher layers. Above them, the higher-layer stack, also know as the Bluetooth host, is responsible for providing the interaction between the applications and the controller. The key components of the higher-layer stack are L2CAP, SDP and GAP. They are fundamental to all of the profiles and transports that serve them.

4.7.1 Logical link control and adaptation protocol (L2CAP)

We have already encountered L2CAP – the logical link control and adaptation protocol. This is responsible for interfacing all of the data applications that use ACL links. It provides multiplexing between higher-layer protocols, so that multiple applications can share the same lower-layer links. It also provides segmentation and reassembly, so that large application packets can be split and fitted into the limits of the lower-layer packet PDUs.

The L2CAP also provides features to control the quality of service of a connection. Flushing timers and retransmission limits can be set for the ACL channels, as well as enhanced reliability and flow control. These features are normally specified by an application profile, so that designers do not need to bother with setting them.

4.7.2 Service discovery protocol (SDP)

Service discovery is a necessary part of Bluetooth's ad hoc capability. Unlike wired networks (and many wireless networks) where the community of connected devices is largely static, Bluetooth is built on the premise that a device is mobile and may make many ad hoc connections during its life. If this is to work when there is a wide range of possible devices and applications to connect to, it is important that there is a mechanism for devices to discover each other's capabilities.

The SDP portion of the standard defines a server database that exists within each Bluetooth device, which lists everything that a device is capable of doing. These are listed as service records, which take the form of 16-bit attributes. These records are unique UUIDs, which are listed in a central assigned numbers list. The higher-level attributes are defined in a list of assigned numbers, whilst the individual profile-specific ones get defined within the profile specifications.

Each device's SDP server database contains information on all of the profiles and protocols that the device supports.

An associated profile – the service discovery application profile (SDAP), which is sometimes confusingly called the service discovery profile, explains how devices can interrogate each other's service database after an L2CAP link has been established. All Bluetooth devices implement the features of an SDP client as well as having an SDP server database. The SDAP provides a number of options for a client to interrogate the database on another device, once an L2CAP channel has been established. This allows it to browse the entire database, or to issue a query for a specific piece of information, searching for a service either by service class or service attributes. The latter is useful where a simple, low-power device does not want to spend time examining every item within its target's server database.

4.7.3 Generic access profile (GAP)

The generic access profile defines the way in which Bluetooth devices discover each other and make their connections. It is the most basic of Bluetooth profiles, but is used by every other profile as the foundation for establishing the link.

The GAP allows a device to be set into one of the three different discoverable modes – non-discoverable, limited discovery and general discovery. It controls the formation of the connection by governing the use of inquiry and paging, looks after pairing and controls when and how security and encryption will be applied to the link.

It also allows the device to be set in connectable or non-connectable mode. When set in non-connectable mode, a device will reject any attempt by another Bluetooth device to pair with it.

4.7.4 Bonding and pairing

Bonding and pairing were touched on in Chapter 3. They are terms that describe setting up a secure link between two devices. They've also become the terms that cause most consternation and confusion for users. Bonding and pairing should be simple – all of the tools exist within the standard to make them so. However, Bluetooth devices still seem to mystify users at this stage of connection. Designers should try and ensure that their pairing techniques are as simple as possible to avoid confusion. Using secure simple pairing (SSP) not only improves the security level, it provides tools to help make pairing far more straightforward. It should be used for all new designs.

The GAP terminology refers to bonding as, 'The dedicated procedure for performing the first authentication, where a common link key is created and stored for future use.' Pairing is the culmination of this process, where, 'A Bluetooth device has a link key that has been exchanged.' A paired device can optionally be set as a trusted device, where, 'A paired device is explicitly marked as trusted.' For the user, all three portions of the process – discovery, pairing and bonding are normally rolled into one procedure, which is colloquially called pairing.

For most users, the pairing experience is one where a passkey or PIN needs to be entered on one or both devices. In legacy security modes (prior to the secure simple pairing of version 2.1), this PIN was used as part of a seed key for the encryption machines within the devices. The first stage of the security process is authentication (bonding in Bluetooth parlance), where the two devices have established a connection. They then check to see that they each share a secret key. To be secure, no key is sent over the air, as the link is still unencrypted at this stage. Instead the master sends a random number to the slave.

The slave ORs this number with the PIN that the user has entered (or which may be preprogrammed into the device if it has no user interface), encrypts it using its secret key and sends the result back to the master. The master performs the same calculation, using the same inputted PIN and compares this with the result it has received from the slave. If they match, both devices know that they share a secret link key and that they are authenticated to talk to each other. To maximise the security, the PIN should be 16 bytes long and preferably alphanumeric. The shorter it is, the weaker the security.

The GAP allows a device to be set as trusted, in which case it will not ask for credentials on future connections, but will automatically connect the devices.

Secure simple pairing (SSP) is introduced in version 2.1 of the Bluetooth standard. This uses advanced cryptographic algorithms to generate the keys, removing the need for the user to enter a PIN. At the end of the pairing process, SSP can output a six-digit number on both devices. The user can thus confirm that both devices are correctly paired – if they are, then the number displayed will be the same on each. Other alternative confirmations are possible, such as flashing lights or playing an audio sequence.

4.8 Transport protocols

A number of different transport protocols have been specified, which move data between applications and the L2CAP layer. Three main protocols are used – RFCOMM, MCAP and AVDTP. Two profiles – HID and AVDTP – include their own protocol transports and interact directly with L2CAP.

The RFCOMM (RF communication – although it is never called that) protocol is by far the most widely used protocol and is derived from the GSM 07.10 serial multiplexing protocol. It is essentially a serial port emulation that harks back to the days of RS-232. It allows both data and commands to be sent from a higher layer profile to the L2CAP layer.

The AVDTP is the audio video data transport protocol, which is used for streaming encoded audio and video over an ACL link.

The MCAP is the multi-channel adaptation protocol. It sits under the health device profile and enables multiple, robust data channels that are used for transferring data in accordance with the IEEE 20601 transport protocol for medical devices.

4.9 Profiles

With the underlying layers in place we can move to profiles. There are currently over 25 different Bluetooth application profiles. In some cases, such as printing, there are several profiles covering the same application. This situation evolved as a result of parallel development within different industry interest groups.

Most profiles are very specific to a particular application. For the majority of general-purpose applications, just a few of these are likely to be used. In this section we'll look at the ones that are most applicable for general short-range wireless products. A full set of profile specifications can be downloaded from the Bluetooth website.

4.9.1 Serial port profile (SPP)

The serial port profile has been responsible for the deployment of more different applications than any other profile, ranging from toys to snow ploughs, from credit card readers to milking machines. Its limitation is that none of these are interoperable with each other.

It is the simplest of the profiles, doing nothing more than emulating an RS-232 serial port on top of RFCOMM. It is the most easily understood by design engineers and is a quick and simple way to add wireless connectivity to a device that has an existing serial port. However, its simplicity is partly a reflection of the fact that it defines neither the data protocol nor the data formats passed over it. The effect of this is that there is no higher-level application interoperability within SPP implementations. Designers need to define their own protocols and formats.

Most module and chip vendors provide firmware that supports SPP, allowing it to be configured as a virtual serial cable. It requires minimal understanding of the Bluetooth standard, which accounts in large part for its popularity.

4.9.2 Handsfree profile (HFP)

In its early days, Bluetooth introduced a headset profile (HSP), which provided a simple connection to a headset. This has now been superseded by the handsfree profile, which adds considerable extra control functionality.

The profile defines two different roles – that of the handsfree (HF) and the audio gateway (AG). The audio gateway is normally the mobile phone or a car kit – providing the connection to the remote source of the voice data.

The handsfree profile mandates the use of the CVSD codec for voice data transported using this profile. It also defines a wide range of voice-control features. All Bluetooth application profiles contain a number of different features, some of which are mandatory, some optional and others conditional. As an example, Table 4.3 lists the call control features that are part of the handsfree profile.

4.9.3 Generic object-exchange profile (GOEP/OBEX)

This is the most commonly used profile for exchanging content. It's based on the object exchange profile that was developed by IrDA. Although OBEX is the term that is generally used for this transfer, the Bluetooth standard splits it into a number of profiles that sit on top of RFCOMM.

At the bottom of these is the OBEX protocol, which is essentially a restatement of the IrDA implementation. On top of this is the generic object exchange profile, which provides support for the basic object exchange functions:

- Establish a data exchange session,
- Pushing a data object,
- Pulling a data object.

The GOEP defines the concept of a push client and a push server. Other than these, GOEP needs no knowledge of a device's filing structure, or the content of the files being transferred. The assumption is that the application or the user will take care of directing files to the correct, final destination. This gives OBEX a simplicity of use that has made it very popular. For example, on a camera phone, OBEX push is normally invoked by the user selecting the 'Send' option after taking a photograph. All the user needs to know to

Table 4.3 *Handsfree profile functions*

Feature	Handsfree	Audio gateway
Connection management	Mandatory	Mandatory
Phone status information	Mandatory	Mandatory
Audio connection handling	Mandatory	Mandatory
Accept an incoming voice call	Mandatory	Mandatory
Reject an incoming voice call	Mandatory	Optional
Terminate a call	Mandatory	Mandatory
Audio connection transfer during an ongoing call	Mandatory	Mandatory
Place a call with a phone number supplied by the HF	Optional	Mandatory
Place a call using memory dialling	Optional	Mandatory
Place a call to the last number dialled	Optional	Mandatory
Call waiting notification	Optional	Mandatory
Three-way calling	Optional	Optional
Calling line identification (CLI)	Optional	Mandatory
Echo cancelling (EC) and noise reduction (NR)	Optional	Optional
Voice recognition activation	Optional	Optional
Attach a phone number to a voice tag	Optional	Optional
Ability to transmit DTMF codes	Optional	Mandatory
Remote audio volume control	Optional	Optional
Respond and hold	Optional	Optional
Subscriber number information	Optional	Mandatory
Enhanced call status	Optional	Mandatory

perform this is the target handset or printer, which is found using a Bluetooth device discovery.

The GOEP also supports some other related profiles, which define or contain more detail about a file contents or a directory structure. These include:

- Object push profile (OPP), which defines business card and calendar formats.
- File transfer profile (FTP), which includes details on filing structures in object and target devices,
- Synchronisation profile (SYNCH), which attempts to synchronise directories on the two devices.

None of these are widely or well supported. Unless they are desperately needed, OBEX provides far better usability and interoperability.

4.9.4 Personal area networking profile (PAN)

The PAN profile was introduced to define how two or more Bluetooth devices could connect together to form an ad hoc network and also to connect via an access point to a remote network. It defines three types of network:

- A network access point (NAP), which allows devices to connect to a network router or bridge,
- A group network (GN), which is a self-contained network of multiple devices,
- A personal network (PAN-U) connection.

The PAN profile has been very little used, as its major application for Internet access has been taken over by 802.11. Despite this, it includes a feature that is of interest to some applications, which is the underlying Bluetooth network encapsulation protocol (BNEP). This is a protocol to allow standard Ethernet IP packets to be transported over a Bluetooth link.

4.9.5 Health device profile (HDP)

The health device profile was developed to permit Bluetooth devices to transport data using the IEEE 20601 protocol, which is becoming the standard for interoperable medical and health products. It provides the means to use this protocol over a Bluetooth link, and also requires that the data formats comply with a set of medical-device specialisations defined in a series of IEEE documents within the IEEE 11073 family. The HDP runs on top of the MCAP protocol, which talks directly to L2CAP.

Using these standards, it is possible to use HDP to manufacture medical and health devices which are fully interoperable with each other, in terms of their output data. The Bluetooth SIG developed this profile in response to requirements from the Continua Health Alliance, who produce guidelines for the medical industry. Bluetooth BR/EDR is the first wireless transport option for Continua devices.

The health device profile also includes a novel time-synchronisation scheme, which allows multiple Bluetooth devices to synchronise an internal clock for time-stamping data. Thus measurements from different devices can be synchronised to within a few microseconds of each other. Whilst the primary driver for this feature was the need to time-stamp data from different body-worn sensors, it provides an incredibly accurate time stamp for any distributed wireless application with this requirement.

4.9.6 Human interface device profile (HID)

The HID profile was designed to enable mice and keyboards to use Bluetooth. That application has largely failed to take off, due to stiff competition from low cost proprietary radios. Despite this, HID has become extremely popular for handheld devices and products needing low latency. It currently accounts for the highest level of Bluetooth chip implementations outside handsets, as a result of its use in the Nintendo Wii controller.

The HID profile describes both the profile requirements and also the HID protocol, which interfaces directly to L2CAP. Both have been designed to minimise latency, with the requirement that a Bluetooth HID connection should add no more than 10 ms to the latency of an event.

The HID profile includes a number of schemes to reduce the power consumption of devices that implement this profile, which are covered in Section 4.10

4.9.7 Advanced audio distribution profile (A2DP)

The rather obscurely named A2DP is a profile that enables high-quality mono or stereo music to be streamed to a headset using ACL links. It uses the audio video distribution transport protocol to interact directly with L2CAP.

To start streaming audio, two devices need to set up a streaming connection using the generic audio video distribution profile (GAVDP). During this process the two devices need to negotiate the most suitable application-specific service capabilities. These include the codec that they will use for encoding and decoding the stream and the content protection capabilities.

All A2DP implementations must support a basic codec, known as the sub-band codec (SBC). This is a license-free, low-complexity codec that is part of the A2DP profile. It provides-high quality audio at low bit rates with only low computational complexity. By making it mandatory, it assures a lowest common denominator of performance for stereo headsets.

The SBC is not the most optimised codec on the market, either in terms of compressibility or performance. For this reason, the A2DP profile includes other optional codecs. Currently, these include MPEG-1 and MPEG-2 audio, MPEG-2 AAC and MPEG4 AAC, and Sony's ATRAC codecs. These all attract a license fee, so the decision to include them is down to a manufacturer. A manufacturer may also define and reference other codecs for proprietary implementations. The mandatory SBC codec provides a lowest

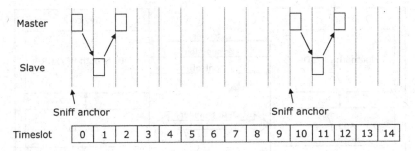

Figure 4.12 Sniff operation

common denominator for occasions where there is no match within the optional supported codecs between the two devices.

A related profile is the audio video remote-control profile (AVRCP), which allows remote control to be added, so that the track, volume and play commands can be controlled wirelessly from the headset.

4.10 Power consumption

By design, Bluetooth is a low-power radio, but for some applications, where data transfer is only occasional, even more power can be saved if the devices are allowed to go to sleep. To accomplish that, the Bluetooth standard incorporates a sniff mode for slave devices. This was enhanced in version 2.1, with the introduction of a sniff sub-rating.

The sniff mode (Fig. 4.12) operates by reducing the duty cycle of a slave device that is in a connection. In normal operation, a slave would listen to traffic from its master on every ACL slot. For many devices, this is wasteful, as there is only intermittent traffic. The sniff mode allows the master and slave to negotiate a reduced number of timeslots in which the slave needs to be awake and listening. These are called sniff anchor points and are defined by the spacing between them – T_{sniff}.

At each of the anchor points, the slave will wake up and listen for a set period to see if any messages are sent to it. If it receives a packet addressed to itself or if it has transmitted an ACL packet, it

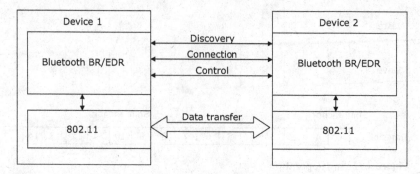

Figure 4.13 Bluetooth 3.0 – the concept

will stay awake until instructed to return to sniff mode. If it does not receive any message it can revert to sleeping after the sniff timeout.

The more aggressive power management system – sniff sub-rating – allows slaves already in sniff mode to skip an agreed number of anchor points. On exiting sniff sub-rating mode, devices will re-enter sniff mode, rather than transitioning directly back to active mode.

4.11 Bluetooth 3.0

The Bluetooth radio is limited in terms of what can be achieved at higher throughputs. However, the well defined protocol stack and application profiles mean that there is a very large installed base of Bluetooth applications that could take advantage of a higher speed transport. Starting in 2005, the Bluetooth Special Interest Group (SIG) started to look at ways of utilising higher-throughput radio technologies to boost the data transfer rate (Fig. 4.13).

The concept behind these was to continue to use Bluetooth BR/EDR to make the connections between devices, set up security and handle data flow but, at the point that higher throughput was required, to direct packets via an alternative high speed MAC/PHY. Effectively, the new MAC/PHY is used as an on-demand, high throughput pipe under Bluetooth control.

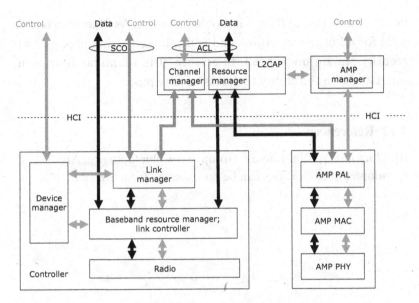

Figure 4.14 Control and data architecture for alternate MAC/PHY

Initial work focused on two alternatives – UWB and 802.11. The UWB development was delayed, as the UWB standard was redesigned to provide a globally acceptable frequency band. Because of this delay, work on UWB was put to one side in favour of using 802.11. The resulting Bluetooth 3.0 standard using 802.11 was released in April 2009.

The key architectural difference is the addition of an alternate MAC/PHY manager (AMP manager), as shown in Fig. 4.14. This integrates with L2CAP and enables multiple alternate MAC/PHYs to be added to the architecture. Each MAC/PHY has its own protocol-adaptation layer (PAL), which provides a standardised interface allowing it to communicate via L2CAP. The Bluetooth higher-layer stack and profiles can then decide when the alternate MAC/PHYs at both ends of the link need to be enabled and data sent over them.

This approach has the advantage that the user interface and user interaction does not change from the one users already know. At no point does the informaiton on which transport is in use need to be communicated to them – it should just work. As Bluetooth is still used for all of the signalling and control, the pairing process and security remain unchanged. In essence, it is a familiar Bluetooth interface, with the ability to access a faster pipe.

4.12 References

[1] Bluetooth Special Interest Group, www.bluetooth.org. All adopted specifications can be downloaded from this site.

5   IEEE 802.11abgn/Wi-Fi

5.1 Introduction

The 802.11 and Wi-Fi standards have become immensely successful in providing Internet connectivity for laptops. In recent years
they have also started to appear in mobile phones and other portable devices to provide a moderate speed connection to Internet
hotspots. They are also finding new uses that take advantage of
the widely deployed infrastructure, notably in the M2M space,
and in some low-power incarnations for asset tracking. The most
recent release of the standard – 802.11n is beginning to garner some
degree of success for audio or video streaming applications in the
home. Despite these uses, almost all current deployments are targeted solely at Internet access.

    802.11 is the oldest of the wireless standards covered in this book.
Its genesis grew out of a proprietary wireless LAN called WaveLAN
that first appeared on the market in 1988, having been started back
in 1986. In its early days, it was not intended for Internet access,
but as a wireless replacement for Ethernet cables, with the potential
markets of factory warehousing and connection to an office network. The concept was to replace the wired physical connection of
the 802.11 standard with a wireless alternative that would slot into
the same 802 protocol stack. In 1991, efforts were begun to evolve it
into a wireless networking standard, which led to the release of the
802.11 specification in 1997.

    As we shall see later in this chapter, much of the topology and
connection states associated with 802.11 result from its genesis as
an alternative connection medium for IEEE 802 systems. This heritage means that it carries a fair degree of functionality that is rarely
used (Fig. 5.1).

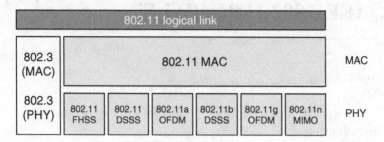

Figure 5.1 802 standards architecture

The first version of the 802.11 standard boasted symbol rates of 1 Mbps and 2 Mbps. Three different wireless interfaces were defined: frequency hopping at 2.4 GHz, DSSS at 2.4 GHz and an infrared connection. Practical data rates rarely exceeded a few hundred kbps. The range of some of the early products was also fairly limited, mainly because of the relative infancy of the technology rather than the specification itself.

The growth in laptops, and the lack of adequate throughput to support Internet access, led to demands to improve the data transfer rate. This initially took place within the 802.11a working group, which proposed increasing the symbol transfer rate to 22 Mbps and moving to the higher ISM band at 5.1 GHz. This proved to be a tough technical challenge for the available technology, so a second working group was formed with a mandate to produce a variant that would be easier to implement within the current technology limitations. This became known as 802.11b. It incorporated an alternative MAC/PHY that operated with a symbol transfer rate of 11 Mbps in the 2.4 GHz ISM band.

A symbol transfer rate of 11 Mbps translates to around 4.5 Mbps of actual Internet throughput (we'll see why later in this chapter), which made it an ideal fit for laptops to connect to the broadband connections that were starting to appear. However, its early years were plagued within interoperability issues, as products from one vendor failed to work with products from another.

This highlighted a difference in specification approach and content that can broadly be divided into 'telecom-oriented specifications' and 'PC-oriented specifications'. This is largely a historic divide. Telecom specifications generally include conformance tests and are accompanied by a certification process requiring products to be tested for conformance before they are allowed onto the market. It's a process that typically slows down the completion of a standard and adds cost to products, but it helps to ensure interoperability, which helps achieve a rapid market acceptance of the technology. In contrast, 'PC-oriented specifications', which is the class to which I would assign the IEEE 802 standards, only define the technology and assume that the market will sort out interoperability. In a competitive market, where companies are trying to differentiate themselves, it runs the risk that they will concentrate on innovation that breaks rather than guarantees interoperability.

5.1.1 The difference between 802.11 and Wi-Fi

This is exactly what happened with the first generation of IEEE 802.11b products. Amidst media reports of security being hacked, deployment stagnated, particularly in the corporate market. To address this, concerned manufacturers formed the Wi-Fi Alliance [1] with the aim of correcting these omissions. The Wi-Fi Alliance revisited the 802.11b specification, correcting errors and inconsistencies and released its Wi-Fi standard along with a test specification and qualification program (Fig. 5.2).

Over the years, the Wi-Fi Alliance has worked in parallel with the IEEE 802 working groups, taking their work and selectively incorporating it into new versions of the Wi-Fi standard. In general, these are compatible with products that conform to the 802.11 version of the standard, although there are features of each which are not included in the other.

The Wi-Fi standards and qualification programs have solved most of the interoperability issues and heralded the start of the

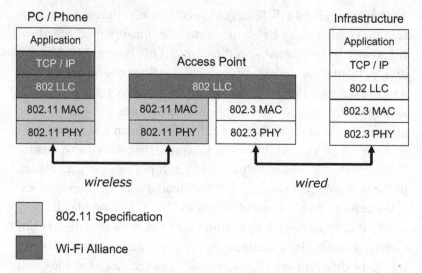

Figure 5.2 802.11, 802.3 and Wi-Fi relationships

success of Wi-Fi. Manufacturers can still design products to the associated 802.11 standards, but these increasingly lack usability functions that the Wi-Fi Alliance has added. Nor is there any guarantee that they will work with a Wi-Fi product. On the other hand, if a product is to be described as 'Wi-Fi' it must pass through the Wi-Fi verification process, with the payment of appropriate membership and qualification fees.

In recent years, the Wi-Fi Alliance has felt it necessary to base its standards on prerelease versions of forthcoming 802.11 standards, as it believes that market demand is moving faster than the 802.11 working groups. This illustrates the different imperatives of an industry-based standards body, which is more reactive to market need, as opposed to an open standards group, which can sometimes debate a standard for years. Table 5.1 gives a brief timeline of the major specification releases from both groups.

All of the increases in speed are actually amendments to the core 802.11 specification. This helps to ensure backwards compatibility, but at times has made it difficult to combine that with optimising the standard, particularly in terms of throughput and power management.

Table 5.1 *Schedule of major releases and features*

Version	Content	Date
802.11	The original standard supporting 1 Mbps and 2 Mbps with both infrared and 2.4 GHz RF physical layer options	1997
802.11a	The 54 Mbps version of the standard, using the 5.1 GHz ISM band	1999
802.11b	Enhancements to 802.11 at 2.4 GHz to support 5.5 Mbps and 11 Mbps	1999
Wi-Fi 'b'	Wi-Fi Alliance certification tests for 802.11b	2000
Wi-Fi 'a'	Wi-Fi Alliance certification tests for 802.11a	2002
802.11g	An enhancement to the 2.4 GHz 802.11b to add OFDM and increase the symbol rate to 54 Mbps	2003
Wi-Fi 'g'	Wi-Fi Alliance certification tests for 802.11g	2003
WPA	Wi-Fi security specification based on 802.11i	2003
WPA2	Wi-Fi Protected Access 2: enhanced security; mandatory for Wi-Fi products shipped after 2006	2004
WMM	Wi-Fi Multimedia: an extension that improves QoS and introduces improved power-saving mechanisms	2004
Wi-Fi 'n'	Wi-Fi Alliance certification tests for 802.11n, based on draft 2.0 of the 802.11 standard	2007
Wi-Fi protected set-up	An optional Wi-Fi standard designed to simplify the user experience of connecting compliant products	2007
BT 3.0	Bluetooth 3.0: uses the 802.11abg MAC/PHY as a high-speed transport for Bluetooth packets	2009
802.11n	A high-throughput specification using MIMO techniques, available for use in either the 2.4 GHz or 5.1 GHz band	2009

A new release of the 802.11 standard, bringing together all of these enhancements up to and including 802.11–2007 was published in July 2007 and provides the best reference to work with.[2] The next cumulative release is currently planned for 2012.

The Wi-Fi Alliance has concentrated on infrastructure modes for their specifications that do not include support for 802.11's ad hoc topology. This should be addressed in a forthcoming release called Wi-Fi Direct, which is scheduled for completion in the second half of 2010.

Today, the Wi-Fi Alliance provides the following certification testing for devices that conform to its standards:

Mandatory

- Core MAC/PHY interoperability over 802.11a, 802.11b, 802.11g and 802.11n,
- Wi-Fi security. This is currently Wireless Protected Access 2 (WPA2) security. Either WPA2 Personal for consumer use, or WPA2 Enterprise, which includes EAP authentication.

Optional

- 802.11d international roaming extensions: these allow products to be made that should operate legally anywhere in the world,
- IEEE 802.11h extensions for 802.11 devices operating at 5.1 GHz (note that 802.11h is mandatory for products that are shipped into certain territories),
- Wi-Fi Multimedia (WMM) quality-of-service and power-saving modes,
- Wi-Fi protected set-up – a specification developed by the Alliance to ease the process of setting up and enabling security protection on small office and consumer Wi-Fi networks.

For much of the time, the technical details of the two standards are the same and products often interoperate. Throughout the rest of this chapter, everything applies equally to both the 802.11 and Wi-Fi Alliance standards, unless it is explicitly stated that it is specific to one of them.

5.1.2 Bluetooth 3.0

Although the assumption made by the IEEE and supported by the Wi-Fi Alliance is that the 802.11 MAC/PHY would use the same UDP and TCP/IP stacks as a wired network, it is not necessary to use these. Recently, Bluetooth has adopted the 802.11 MAC/PHY as a high speed transport for Bluetooth products, using Bluetooth's higher-level protocol stack. For more information on this, see Chapter 4.

5.1.3 Alphabet soup

All of the working groups within the 802.11 standard are characterised with a letter suffix. As the standards have evolved, these have multiplied. As well as the headline groups of b, a, g and n, which have worked on progressively higher throughput physical layers, Table 5.2 lists those more relevant to implementers.

A full list of the working groups can be found on the IEEE 802.11 website.[5] There is also a good overview on the Wikipedia page.[6] To avoid letter and number confusion there is no 802.11o or 802.11l.

5.2 802.11 topology

802.11 supports two different topologies – ad hoc and infrastructure. Of these, almost all products use a variant of the infrastructure mode, generally as a subset of its full capabilities, where an access point is only connected to a single back-end broadband link. The heritage of 802 networking means that there is a lot more functionality available. That heritage also governs the topology and much of the nomenclature around 802.11.

All 802.11 networks start with a device that acts as the central point to which other devices connect. In the case of infrastructure

Table 5.2 *Key 802.11 working groups*

Group	Content	Status
802.11d	International (country-to-country) roaming extensions	Released 2001
802.11e	Quality-of-service enhancements	Released 2005
802.11h	Spectrum management (power control and dynamic frequency selection) for the 5 GHz band in the European market	Released 2004
802.11i	Enhanced security: a basis for the WPA security schemes	Released 2004
802.11j	Enhancements to cover regulatory requirements in Japan	Released 2004
802.11p	A variant of the standard, running in the 5.8 and 5.9 GHz bands for vehicle-to-vehicle and vehicle-to-infrastructure applications. Also known as DSRC (dedicated short-range communications),[3] which includes WAVE (wireless access for the vehicular environment). In Europe, this is being developed by the Car2Car Consortium.[4]	Ongoing
802.11r	Fast roaming: an initiative that will bring a cellular-type roaming experience to 802.11	Ongoing
802.11s	Mesh networking for 802.11	Ongoing
802.11u	Internetworking issues with non-802 networks: this group is looking at how 802.11 will coexist with other wireless technologies, in particular with cellular	Ongoing
802.11aa	Improving audio and video streaming performance	Ongoing
802.11ad	A group looking at very high throughputs in the 60 GHz band, currently used by WirelessHD	Ongoing

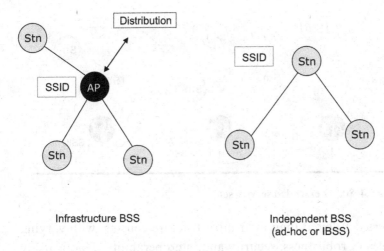

Figure 5.3 802.11 topologies

mode, this is the access point. For ad hoc networks, it is the initiator of the network. 802.11 calls these individual networks base service sets (BSS). The fundamental characteristic of a BSS is that it has a service set ID, more commonly known as its SSID. The SSID is an alphanumeric string of 1 to 24 characters, which identifies the node in the network to which all other nodes connect. (There is a special zero-length SSID, which is used for probe requests, when a station wishes to find all of the available access points.)

Figure 5.3 illustrates the two basic topologies that can be formed around a base service set. In the case of an infrastructure connection, the access point announces its SSID for other nodes that wish to join its BSS. Access points also connect to a distribution function, typically a broadband link. This is known as an infrastructure base service set.

An ad hoc connection is essentially the same, except that the node advertising its SSID does not have any further external connection. Ad hoc networks are confusingly known as independent base service sets or an IBSS. (Note that this acronym is never applied to the infrastructure base service set.) Ad hoc connections are not currently covered by the Wi-Fi certification scheme, and

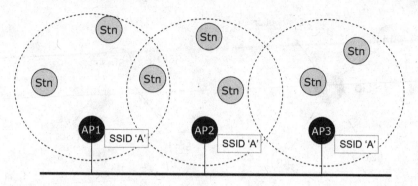

Figure 5.4 802.11 extended service set

are less developed than their infrastructure cousins, with varying degrees of robustness, security and interoperability. That is likely to be corrected during the course of 2010, when the Wi-Fi Alliance delivers its Wi-Fi direct standard.[7]

The networking background of 802.11 assumed that the most common use would be to cover corporate and campus environments where users would move between access points, but would expect to have a continuous connection without the need for users to do anything. This is made possible by the concept of the extended service set (ESS).

Figure 5.4 shows this arrangement, where several access points are connected to the same backbone. Typically this is a wired Ethernet backbone, but it could be any other connection. If all of the access points on the backbone have the same SSID, then nodes that move from the coverage of one access point to another are able to disassociate and reassociate with the new access point, without any break in their network session. This assumes that they have a strong enough signal to maintain a connection during that process.

Figure 5.5 shows how access points with different SSIDs can share the same backbone, but implement distinct extended service sets. In this case, stations connected to AP1 could roam to the area covered by AP2, but would not be able to associate seamlessly with AP3 or AP4.

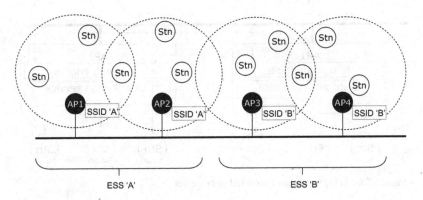

Figure 5.5 802.11 multiple extended service sets

An access point can select to hide its SSID, so that it is not included in beacons that it sends out. This is sometimes encouraged as a security feature. However, a station that knows its SSID (as it would have to if it wanted to connect to it) still sends out packets with the SSID unencrypted, so an attacker could easily discover the hidden SSID, hence the level of security provided is minimal.

5.2.1 Bridging with access points

To allow roaming between access points within an ESS, access points need to implement a bridging function, which keeps a record of which stations are attached to which access points. The concept is illustrated in Fig. 5.6.

Each access point in an ESS needs to keep information on all of the nodes connected to it, and share this information with every other access point within the same ESS. When a message is transmitted from a node, the access point will use the bridging engine to forward it to the appropriate access point.

This allows stations to use association functions to manage their movement between the access point BSSs within an ESS. The process is shown in Fig. 5.7. Movement from one access point to another is always initiated by the individual station, never by the access point.

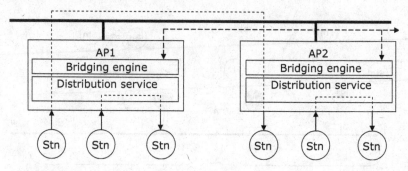

Figure 5.6 Bridging in an extended service set

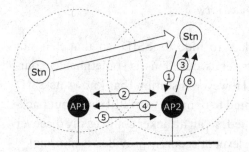

Figure 5.7 Moving around an extended service set

As the station moves out of range of the first access point (AP1), it discovers a new access point within the same ESS (i.e., which has the same SSID) – in this case AP2. To change association, the following sequence is followed. I'll explain the details of some of these functions in the next section.

1. The station sends a reassociation request to AP2, which includes the physical address of the old access point – AP1.
2. The new access point uses this to check with AP1 to ensure that the station had a valid association with it.
3. If this confirms that there was a valid association with AP1, AP2 informs the station that it is now associated with itself (AP2), and the two access points update their bridging tables.
4. AP2 can optionally ask the original access point whether there are any buffered data for the station. This could be the case if

the station was in a low-power mode and had moved location before waking.
5. If any data are cached by AP1, they are transmitted to AP2.
6. AP2 now transmits this on to the station.

At no point in the roaming or reassociation sequence does the station communicate with the first access point. It could disassociate or deauthenticate with it prior to making a connection with AP2, but this would result in the network session being terminated.

Where 802.11 is deployed with a backbone providing a distribution function across multiple access points, stations can seamlessly roam without the need to terminate and restart their higher-layer network session. At the start of its deployment, it was envisaged that the primary use for 802.11 networks would be for corporate networks, where extended service sets would allow employees with portable devices to move freely around the building. This stalled when the first security issues with WEP were widely publicised. Concerns about network security saw the corporate market withdraw from 802.11. An unexpected growth in home networking, where a single access point was initially connected to and later incorporated into a broadband modem saved the industry, as consumers took to the technology. Although security concerns have now been addressed, the bulk of access points shipped still have a single connection to a broadband link, with roaming capability used only on a minority of access points.

5.2.2 802.11 services

The topologies and roaming features described previously are enabled by a core set of features that were incorporated within the first version of the 802.11 standard and have remained in place for all subsequent versions. These are split into station services and distribution services. Station services cover the connection of devices, whilst distribution services support the movement of data around the network.

5.2.2.1 Station services

The four station services are authentication, deauthentication, encryption and MSDU delivery (MAC service data unit). They explain how a station connects to another device and sends data securely. The station services are used both for ad hoc and infrastructure network connections.

Authentication is used to establish the credentials of a station at the point where it attempts to make a connection. It involves a simple exchange of the MAC addresses of the two devices and is a necessary precursor to association.

The use of the term 'authentication' within 802.11 can be confusing. The original meaning of this service, as explained above, is a very simple exchange of information with another node. It does not involve the subsequent security authentication that will normally follow on after a successful association (q.v.).

Deauthentication is the process of terminating an authenticated relationship. It is important in secure networks, as deauthentication is responsible for clearing any stored keys related to that connection. As authentication is the lowest service within a network connection, performing a deauthentication will also terminate any association if it is performed without a prior disassociation.

Encryption, or **confidentiality**, covers the services that prevent eavesdroppers from discovering the contents of the data payload. As the standard has been tested in the real world, this has evolved from WEP through WAP, WPA, 802.11i to WPA2. Some of these more recent services are not currently applicable within ad hoc connections.

MSDU delivery: MSDU is the MAC service data unit, which has the task of transferring data from a transmitting device through to its endpoint on the network.

5.2.2.2 Distribution services

There are five distribution services, which are employed to move data around the network and to integrate the wireless network into a larger network structure. They are association, disassociation, reassociation, distribution and integration.

Association follows on from authentication and 'registers' a station with an access point or ad hoc master. This registration is taken further in a distributed system, where it is passed on by the distribution function to other access points to inform them of the route to access the node. It can be considered as a method for access points in a distributed system to manage a distributed database of who is where.

Disassociation is the reverse of the process, where a station from the network removes itself from an access point. Disassociation allows the access point to remove its association data and to transmit this change to any access points that had knowledge of this through the distribution function.

Disassociation should be performed prior to removing a station, either as a result of poor signal strength, or by shutting it down. As a formal disassociation cannot be guaranteed under all operating conditions, the MAC of 802.11 is designed to cope with devices that 'spontaneously disappear'. From a network management viewpoint, it is good practice to try and dissociate wherever possible.

Reassociation is used when a station wishes to move to a new access point within the extended service set. Reassociation is always initiated by the mobile station. The conditions for reassociation are not defined within the standard, but are down to manufacturer implementations. Normally this would be triggered by sensing a falling RSSI signal from the currently associated access point, accompanied by the presence of an alternative access point within the ESS. However, some access points may deliberately remove stations, forcing them to attempt to reassociate with another access point.

Distribution is the service that takes care of delivering each frame of data. Every frame holding data from a station also contains the destination address for the data. It is the role of the distribution service to ensure that the data reach their destination. The distribution service covers data directed between two stations attached to the same network, as well as data transferred to the wider network.

Integration is the final distribution service. It is provided to cope with networks where the 802.11 wireless network does not connect through another 802.11 network. It provides a set of services that allow integration of the 802.11 distribution service with other services. It does not extend to the specifics of other networks.

Two other services, deriving from 802.11h, are sometimes added to the service set. These relate to the way in which radios operating in the 5.1 GHz band need to behave to meet European requirements.

In Europe, the 5.1 GHz band had been reserved for an alternative wireless network standard known as HiperLAN (high performance LAN). This standard was overtaken by 802.11a, but the European regulatory authorities imposed some behavioural requirements on any radios operating in this band to try to minimise interference, particularly with frequencies that are also used for radars. The 802.11a specification does not include these features, and they have been added in the 802.11h specification. These need to be implemented in any 5.1 GHz wireless LAN shipping into an EU member country. They represent good practice for any implementation, regardless of its destination.

Transmit power control (TPC) mandates that radios will monitor the condition of the radio link and adjust the transmit power of both devices to a level that is just sufficient for reliable operation. As the receive sensitivity and received signal quality may differ for the two nodes, TPC needs to work independently for the two link directions.

Dynamic frequency selection (DFS) requires that 5.1 GHz wireless LANs periodically listen for other fixed transmitters within the band. If they are discovered, the units should intelligently move the network to a clear channel within the allowed band.

5.3 The 802.11 radio

The original 802.11 specification concentrated on replacing the wired PHY and associated MAC of 802.3 with a wireless equivalent. The initial specification covered three different alternatives: an infrared transceiver, a frequency-hopping radio operating at 2.4 GHz and a direct sequence radio, also operating at 2.4 GHz. The

Table 5.3 *PHY variants and throughput*

Standard	Spectrum (GHz)	Typical throughput (Mbps)	Symbol rate (Mbps)	Typical range (m)
802.11	2.4	~0.8	2	100
802.11a	5.1	~24	54	15
802.11b	2.4	~5	11	45
802.11g	2.4	~22	54	25
802.11n	2.4; 5.1	~130	600	50 (at 2.4 GHz)

industry coalesced around the direct sequence radio and this has become the basis for all future enhancements.

Table 5.3 shows the effect of the PHY evolutions since the first version in 1997. Typical throughputs assume a TCP/IP stack. Throughputs will be slightly higher if a UDP stack is used.

All of these variants use a direct sequence spread spectrum (DSSS) technique for the radio. As 802.11 operates in an unlicensed frequency band, there are requirements that its radios do not monopolise the band. It is also in the interest of each standard that it can cope with interference from other radios.

The DSSS technique works on the principle that much of the noise within a frequency band will come from narrowband transmissions. Rather than trying to avoid these, DSSS uses a spreading function to transform its signal across a wider frequency range within the spectrum (Fig. 5.8). At the receiving end, the reverse transformation takes place in a correlator, reproducing the original signal.

Any noise arriving at the receiver goes through the same reverse transformation, with the result that it is reduced from being of similar amplitude to the received signal, to an order of magnitude or more lower (Fig. 5.9). This process effectively improves the signal-to-noise ratio.

One consequence of spreading the signal is that the energy of the transmitted signal typically expands to fill more of the channel.

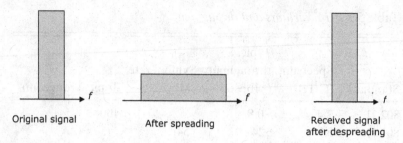

Original signal   After spreading   Received signal
                                                  after despreading

Figure 5.8 How DSSS works

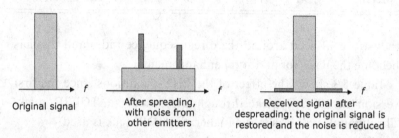

Original signal   After spreading,   Received signal after
                   with noise from   despreading: the original signal is
                   other emitters   restored and the noise is reduced

Figure 5.9 Noise rejection with DSSS

To control this, transmitters often include some degree of filtration and signal shaping. As the modulation schemes for coding the data become more complex, this further broadens the energy profile. When high-power transmitters are used, it can be difficult to contain these within the regulatory limits.

As we have seen before, there is always a trade-off in wireless. The DSSS transmitters do not hop, but work at a single fixed frequency. Their receivers exhibit a better rejection of interference, but require more bandwidth, so DSSS systems invariably have fewer channels in the same chunk of spectrum than do frequency hopping networks.

In the 2.4 GHz band, where most of the 802.11 products reside, these channels are 22 MHz wide, spaced 5 MHz apart. They are numbered from 1 to 13, starting at 2.412 GHz and ending at 2.472 GHz. Japan allows an additional 14th channel, at 2.484 MHz.

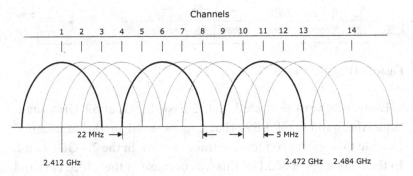

Figure 5.10 2.4 GHz spectrum usage

The global availability of channels is not standardised. Japan allows all 14 channels, although channel 14 cannot be used if the radio is using orthogonal frequency-division multiplexing (OFDM). Europe allows bands 1 to 13, although France limits the power to 10 mW for use outside buildings. The USA only allows channels 1 to 11, although powers up to 1 W may be used with certain restrictions.

This channel width of 22 MHz and spacing of 5 MHz means that there is considerable overlap between the channels, as shown in Fig. 5.10. As can be seen, in the USA, which is limited to channels 1 to 11, there are only three non-overlapping channels: 1, 6 and 11. The heavier lines indicate these. Most commercial products will ship with the channel preset to one of these. Europe allows four non-overlapping channels – 1, 5, 9 and 13, although it is still common to see most products shipped with the defaults set to either 6 or 11.

In the 5.1 GHz band, there is a similar core of 12 or 13 channels available around the world, but with the major advantage that the channel spacing is 20 MHz, instead of the 5 MHz within the 2.4 GHz band. This means that channels no longer overlap, allowing use of all of these channels. The channels from 5.180 (channel 36) to 5.320 GHz (channel 64) are contiguous. In many countries there is then a break, up to 5.745 (channel 149) with 20 MHz spacing going from there to 5.805 (channel 161). Increasingly, use is being allowed across this whole band, but there remain differences from country to country.

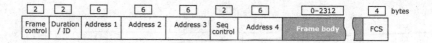

Figure 5.11 802.11 frames

Even at its most restricted, the 12 basic bands at 5.1 GHz quadruple the number of BSSs that can be used in the same area. Despite this, most products continue to work in the 2.4 GHz band. In the early days of 802.11a, this was because of the added cost and technical difficulty of operation at 5.1 GHz. Now, it is more likely because of compatibility with an installed base of 2.4 GHz access points, although dual-mode designs capable of operating at either frequency are beginning to increase the infrastructure. There is a wider variation in national regulations in this band, although this is a core spectrum, which is available throughout most of the world.

One other band, which is used for a vehicle-specific variant, is defined by 802.11p. This is situated in the 5.8 and 5.9 GHz spectrum and is not yet globally harmonised. This application is currently not addressed by the Wi-Fi Alliance, but is being developed by a variety of consortia within the automotive industry.

5.4 Framing

Although any detailed discussion of framing is beyond the limits of this chapter, a brief overview of 802.11 frames is useful to help understand the way in which different coding schemes are implemented and why there is variation in the throughput. For further details of coding, I recommend Matthew Gast's excellent book.[8]

The generic frame for 802.11 transmissions is shown in Fig. 5.11. Anyone familiar with Ethernet frames will recognise key features of it, in particular the multiple address fields. As with every short-range radio and every device conforming to an 802 standard, each device has a unique 48-bit address. The first two address fields within the 802.11 frame include the address of the final destination

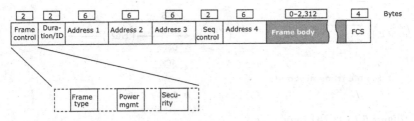

Figure 5.12 802.11 control frames

(address 1) and the sending device's address (address 2). The other two fields may contain other addresses, or additional information, depending on the type of frame.

The frame type is determined by the first segment of the generic frame – the frame control (Fig. 5.12). Amongst many other things, this section of the generic frame identifies the type of frame, whether and what type of power management is being used and the security level.

The bulk of any generic frame is the frame body or data field, which contains the data being carried by the generic frame. This can contain up to 2312 bytes of information. On an IP network, which is the normal case for 802.11 usage, this is limited to 1500 bytes, which may be further reduced to around 1400 bytes when connecting to a DSL network.

At the end of the frame, a 4-byte frame check sequence is applied to the contents of the rest of the frame. This is checked for each incoming frame to determine whether or not it is corrupted. Unlike wired Ethernet, 802.11 does not contain any form of acknowledgement for a bad frame. If a frame arrives and fails the frame-check sequence, then the sender needs to wait for an acknowledgement (ACK) timeout, at which point it will resend the entire frame. This can have a significant effect on throughput when a connection is operating close to the extreme of its working range.

One other important part of the generic frame is the duration segment. Amongst other information, it includes the value of the network allocation vector (NAV). This is a key component of managing access to the wireless media.

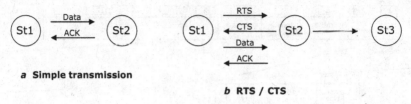

a **Simple transmission**

b **RTS / CTS**

Figure 5.13 802.11 radio access

To understand NAV, we need to look at how 802.11 radios cope with accessing the radio medium without interfering with each other. If every station were able to transmit at will, then, as the network density increased, there would be an increasing number of collisions, with a resulting loss of throughput. To guard against this, 802.11 uses a distribution coordination function (DCF) to try and minimise the possibility of collisions. Figure 5.13 shows how this works.

Before transmitting, stations need to listen for a short period to check that nothing else is using the network. Short frames can just be transmitted in the expectation that they will get through. The transmitter then waits for an acknowledgement (Fig. 5.13a) and tries again if none arrives. For a very lightly loaded network with few stations, this may not be a problem. However, listening for traffic may not be enough. Figure 5.13b shows Station 1, which wants to transmit data to Station 2. Although it may be listening to see if anyone else is using the network, it is not aware of Station 3, as this is out of its range. However, if it were to transmit at the same time as Station 3, the frames from these two stations would collide at Station 2, which is able to hear both.

To overcome this, an RTS/CTS scheme can be added. Using this, Station 1 listens to ensure nothing else is transmitting, and then sends a short RTS frame to Station 2. Station 3 will not hear this, but it will hear the CTS response that Station 2 sends back to Station 1 and as a result knows that it must not transmit. When Station 1 receives this frame, it knows that the network is clear and can transmit its data frame to Station 2.

The question this poses is how does Station 3 know how long it must keep quiet for? This is where the DCF comes in. It provides

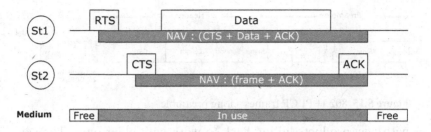

Use of NAV to signal wireless medium availability

Figure 5.14 802.11 network allocation vector

an indication of network availability by setting a time value in the duration segment of each frame. This is the network allocation vector (Fig. 5.14). The NAV is essentially a statement of the amount of time that Station 1 needs to complete its entire transaction of RTS/CTS/Data/ACK. Every station hearing this will set its NAV timer to this value and then count down, knowing that it is not allowed to transmit until its internal NAV counter reaches zero.

Station 3 will not initially be aware of this, as it is out of range of Station 1. To get over this, when Station 2 returns its CTS packet, it includes a recalculated value of NAV that covers the remainder of the transmission. Station 3 hears this and can therefore set its own NAV counter.

All devices listen for these NAV values. As each transmission includes a NAV value, this ensures that any station within range of either station involved in a transmission will know the duration for which these stations need clear access to the network.

The DCF is the simplest and most common of a number of coordination functions that are provided for access to the 802.11 network.

5.5 Modulation

As each new version of the standard has appeared, data transfer rates have increased. Despite this, 802.11 products working in the same frequency band have maintained backwards compatibility,

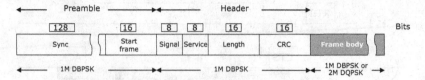

Figure 5.15 802.11 PLCP frames – long preamble

with new products falling back to work with older ones. The key to this has been the way in which 802.11 has addressed the issue of packet modulation in its successful attempts to cram an ever increasing amount of data into the same frame.

The first release of 802.11 used two coding schemes, both employing differential phase shift keying (DPSK), where the data are encoded as phase changes in the transmitted signal. The simplest, providing one symbol per bit, is differential binary phase shift keying (DBPSK), which gives a headline symbol transfer rate of 1 Mbps. The second, differential quadrature phase shift keying (DQPSK) encodes twice as many symbols per bit, doubling the amount of data in the data frame. Stations and access points negotiate the rate at which they will operate, choosing the highest reliable rate. The overheads surrounding the data payload, coupled to media access constraints, are the reason that the maximum throughputs of 802.11 devices are typically just under half of the headline symbol rate.

The frame generated in the MAC layer that we looked at above, is packaged using a physical layer convergence procedure (PLCP) into a frame that is passed on for transmission. This includes the addition of a preamble sequence, which is used by the receiving radios to synchronise and align the frame, and a header sequence, which provides details about the frame. The signal segment of the header identifies the encoding of the MAC frame.

As Fig. 5.15 shows, where DQPSK is employed, this is only applied to the MAC data frame. The preamble and header are still encoded in the 1 Mbps DBPSK form. This means the halving of the size of the data packet resulting from DQPSK encoding is not realised for the preamble and header, but is confined to the payload data.

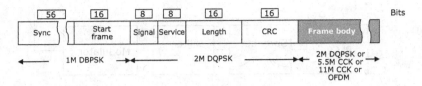

Figure 5.16 802.11 PLCP frames – short preamble

802.11b made a major step forward in throughput by introducing new coding schemes that pushed the symbol rate up to 11 Mbps. This is known as the high rate PHY. Two different schemes are specified in the standard – complementary code keying (CCK) and packet binding convolutional coding (PBCC), but only CCK is widely used. All that will be said here is that CCK stuffs more data bits into each symbol, providing new rates of 5.5 Mbps and 11 Mbps.

Again, this coding is only applied to the contents of the MAC frame. As a result, the overhead of the preamble starts to have a more serious impact on throughput. To help reduce this, the 802.11b standard introduced the concept of short preambles, where the 144-bit preamble is halved to 72 bits and the following header section is encoded as 2 Mbps DQPSK (Fig. 5.16).

This short preamble is used again in 802.11g, which added orthogonal frequency division multiplexing (OFDM) encoding to increase the coding rates up to 54 Mbps with the introduction of the extended rate PHY (ERP). This has fallback rates of 6, 9, 12, 18, 24, 36 and 48 Mbps, of which 6, 12 and 24 Mbps are mandatory. The coding schemes are taken directly from 802.11a.

The way in which this is treated by the radio transmitter is illustrated in Fig. 5.17. After being constructed, the PLCP is passed to the DSSS spreader, then through a filter and finally modulated according to the encoding scheme. For short preambles, different modulations are applied to the preamble and header.

As well as DSS coding with a short header, 802.11g also allows more compact OFDM frames. However, these cannot be read by 802.11b stations, which means that they are unable to extract NAV

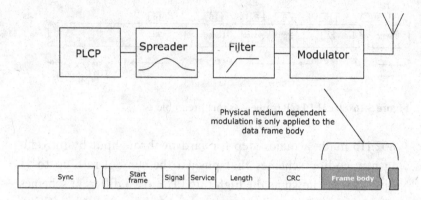

Figure 5.17 802.11 PHY

information from them. Hence, despite the fact they are more efficient, they can only be utilised in a BSS in which all of the stations are running 802.11g.

802.11g is backwardly compatible with all of the coding schemes of 802.11 and 802.11b. However, this imposes a requirement on access points that they support both long and short preambles. Depending on implementation, there have been examples of access points where this is not the case. It may be that a single station using long preambles will force these onto all other stations. More recently some access points have emerged on the market that only appear to support short preambles. These can be problematic when used with some of the ultra-low power 802.11 chipsets, which conserve power by limiting coding schemes to the simpler 1 and 2 Mbps options. If problems are encountered where a station is unable to associate with an access point, or the throughputs are unusually low, this is a good point to start your investigation.

5.6 5.1 GHz – 802.11a

802.11a is very similar in terms of encoding to 802.11g, which is not surprising as it was developed first and provided much of the content for 802.11g. The key difference is that it operates in the U-nii band at 5.18 GHz.

Despite having much more spectrum and no appreciable interference, 802.11a has been slow to gain popularity. This is largely a result of the higher cost of producing 5 GHz transceivers, and the attraction of a large and increasingly widespread installed base of 2.4 GHz access points and public hotspots. That is starting to change, not least because of the availability of reasonably priced dual-band 802.11a+g chipsets, which are capable of operating at either of the two frequencies.

5.7 MIMO – 802.11n

The higher coding schemes of 802.11g exhibit the expected behaviour of greatly reduced range (as shown in Fig. 2.6). To increase rates still further without diminishing range to an unacceptable level, the 802.11 group needed to turn to a different technique – that of MIMO, or multiple input, multiple output.

MIMO refers to a technique of increasing the number of antennae on each device, each of which has its own transmit or receive circuitry. (This is in contrast to antenna diversity, where two antennae are used on a receiver, but the signal is only taken from the one with the best reception). The feeds to or from these transmitters and receivers are aggregated into a number of spatial streams. At the MAC level, frames can be aggregated and combined into multiple streams.

This allows a high-speed MAC to generate streams that are sent from multiple transmitters, each with its own antenna, to corresponding receivers. The nomenclature for this type of MIMO system is:

(number of transmit antennae) × (number of receive antennae):
spatial streams.

Table 5.4 *MIMO throughputs (Mbps)*

Streams	Modulation	20 MHz		40 MHz	
		Long GI	Short GI	Long GI	Short GI
1	BPSK	6.5	7.2	13.5	15.0
2	BPSK	13.0	14.4	27.0	30.0
1	QPSK	19.5	21.7	40.5	45.0
2	QPSK	39.0	43.3	81.0	90.0
1	16-QAM	39.0	43.3	81.0	90.0
2	16-QAM	52.0	57.8	81.0	90.0
1	64-QAM	65.0	72.0	135.0	150.0
2	64-QAM	130.0	144.4	270.0	300.0
4	64-QAM	260.0	288.9	540.0	600.0

The most common implementation is 2 × 2:2. In other words, the transmitter combines two streams to be sent from two antennae, which are received by a device with two antennae and receivers. The configurations at transmitter and receiver do not need to be the same. Having more receiver antennae may improve performance, as it results in an improvement in the signal-to-noise ratio. For example a 2 × 3 configuration has around a 20% higher throughput compared with a 2 × 2. When first connected, devices will negotiate the optimal configuration to use.

To be effective, particularly at the receiver, there needs to be a spacing of several centimetres between antennae. That means that small devices, such as mobile phones, will probably only implement 802.11n as 1 × 1:1

As well as introducing multiple antennae, 802.11n also allows the channel width to be increased to 40 MHz, double that of the other 802.11 specifications and to reduce the guard interval (GI) between frames. These changes result in some impressive maximum throughputs, as shown in Table 5.4.

The high (theoretical) values at the bottom of the table are impressive, but are not representative of most commercial devices.

Implementing four separate transmitters and receivers, with the resulting eight antennae on a device, is expensive and takes up space. In many parts of the world, the 40-GHz-wide channel is not allowed, and if implemented in the 2.4 GHz spectrum will almost certainly run into interference issues. Moreover, these figures are only valid for 'greenfield' sites, where 802.11n is the only 802.11 variant using the spectrum. As soon as other 802.11 systems wish to connect, 802.11n needs to move to compatible PLCP frames, which seriously degrades the throughput. Nevertheless, systems with a 20-MHz channel and a 2 × 2 configuration can achieve throughputs approaching 100 Mbps. An important bonus is that the use of multiple antennae helps to reduce the effects of multipath interference, giving a greater range than would be expected for the output power.

5.8 Making connections

To form a network, 802.11 defines the way in which devices find each other. The starting assumption is that there will be a device containing an SSID, which wants to form the BSS. This may be an access point operating in infrastructure mode, or one or more stations configured to be part of an ad hoc network.

The connection process is always initiated by the station wanting to join the BSS. This can happen in two ways, which are largely determined by the power-saving mode of the joining station.

Active scanning involves the station sending out a probe request. To do this, the station sends a request frame on each of the available 802.11 channels. The request may either be for a specific SSID, or can use a broadcast SSID to elicit a response from every BSS within range. At the end of this scan procedure, the initiating station will have a list of all of the BSSs within range, along with their parameters. It can then decide which to connect to.

Passive scanning does not require any transmission to be made by the joining station, which saves power. Instead it listens for beacons to be sent from the access point or ad hoc master (only one station within a BSS is allowed to send beacons). The limitation is that the beacon interval may be long, so that a station needs to listen for an extended period. As beacons contain all of the information necessary to start a connection, a device that has utilised passive scanning can start its connection procedure immediately after completing its scan period.

5.9 Power management

Because 802.11 networks require a station's receiver to listen for all incoming packets, they consume power whenever they are connected. To reduce this, a number of power-saving schemes have been defined that allow a station to go to sleep. These are all controlled by the access point. This has two implications:

- An access point does not itself have a low-power mode, and
- Low-power modes in stations depend on efficient power management implementations within an access point.

Access points can efficiently manage the power of attached stations by the use of beacons. The access point tells stations when it will send a beacon, and requires the station to wake at certain beacon intervals. Up until that point, a station is allowed to go to sleep. If a frame arrives for a station whilst it is asleep, it will be stored by the access point. When it next sends a beacon, it will include a traffic indication map (TIM) within the beacon, which lists all of the stations that have frames waiting for them. If a station sees its address within the TIM, then it must stay awake to retrieve and act on that frame. Otherwise it can revert back to sleep. A number of different modes are defined, which determine how often a station needs to wake and whether it can access deep-sleep modes.

Access point manufacturers need to consider a range of compromises. The bigger the memory buffer within the access point, the more data it can store for its connected stations. This means that the stations need to wake up less frequently and will consume less power. The corollary is that however good the power management on a station, its ultimate performance is determined by the access point, which is often unknown.

Power management in ad hoc networks is limited as it requires stations to have more knowledge of when the intended recipient is awake. This means that receivers need to be on for longer and there is no equivalent method of allowing deep sleep. As almost all 802.11 developments have been led by infrastructure mode applications, ad hoc has lagged behind in this area. This may well be corrected in Wi-Fi Direct, which promises to bring advanced power saving to ad hoc networks.

5.9.1 Wireless multimedia power save

The Wi-Fi Alliance has extended the power-save features of the 802.11 specification with their wireless multimedia (WMM) power-save extension. Within 802.11, the assumption is that the device driver sitting above the MAC will be responsible for placing the device into a power-save mode. The WMM power-save extension moves the intelligence for power saving further up the stack, so that an application can dictate when and for how long the device can go to sleep. Particularly in cases where they are multiple applications using the wireless link, this provides better granularity of control, allowing the device to doze for longer periods. The Wi-Fi Alliance reports power savings of between 15% and 40% when WMM power save is used.

5.10 References

[1] The Wi-Fi Alliance, www.wi-fi.org.
[2] IEEE Standards Association, IEEE 802.11LAN/MAN wireless LANS. http://standards.ieee.org/getieee802/802.11.html.
[3] IEEE 1609 Working Group, DSRC & P1609 project page. http://vii.path.berkeley.edu/1609_wave/.
[4] Car2Car Consortium, www.car-to-car.org/.

[5]　IEEE 802.11 Working Group, http://grouper.ieee.org/ groups/802/11/.

[6]　Wikipedia, IEEE 802.11. http://en.wikipedia.org/wiki/802.11.

[7]　Wi-Fi Alliance, Wi-Fi Alliance announces groundbreaking specification to support direct Wi-Fi connections between devices. www.wi-fi.org/news_articles. php?f=media_news&news_id=909.

[8]　Matthew S. Gast, *802.11 Wireless Networks: The Definitive Guide*, 2nd edn (O'Reilly Media, Inc., 2005).

6 IEEE 802.15.4, ZigBee PRO, RF4CE, 6LoWPAN and WirelessHART

802.15.4 and ZigBee have become largely synonymous in the minds of many people. That's in large part a result of an excellent marketing campaign by the ZigBee Alliance.[1] In fact, the two are used together to form ZigBee products; IEEE 802.15.4 [2] defines a low-power radio and media access controller (MAC) and the ZigBee Alliance defines a mesh networking stack that sits on top of the 802.15.4 standard.

Although by far the best-known higher-layer protocol stack using the 802.15.4 radio, ZigBee is by no means the only one. There are at least a dozen other standards making use of this low-power radio, of which those with the largest market usage are probably RF4CE,[3] WirelessHART [4] and 6LoWPAN.[5] In this chapter, I'll concentrate on the underlying 802.15.4 standard and ZigBee – in particular the ZigBee PRO standard, but I'll also provide a brief overview of these other three upcoming specifications.

One of the reasons that the 802.15.4 radio is so well known is that there are no licence fees or restrictions around using it. That has made it a favourite for universities and companies developing a myriad of different low-power sensor networks. There is an associated risk, in that there is no guarantee that using it does not infringe patents, but its relative simplicity and the availability of chips and development kits from a number of different suppliers means that it is likely to remain a popular choice. The patent issue is an important one if you're planning to make a commercial product. For an explanation of that, see Chapter 10.

Higher-layer standards like ZigBee bring interoperability to low-power wireless. For any market to grow it's important for products from different suppliers to work with one another. ZigBee has developed a comprehensive application framework and profiles which do exactly that. Without it, 802.15.4 is not significantly different from a proprietary radio.

Like all of the established wireless standards, both 802.15.4 and ZigBee have evolved through a number of different versions, as shown in Table 6.1.

The two groups have different agendas. That of 802.15.4 is to continue to define a series of lower-layer enabling blocks, which can be used for low-cost, low-speed communication between low-power devices. This covers a wide range of potential applications. ZigBee concentrates on developing robust, secure and easy-to-install mesh networks. It is based on the original release of the 802.15.4 specification, which is adequate to support the ZigBee stack. It has not found a need to embrace any of the more recent additions to the 802.15.4 standard.

6.1 IEEE 802.15.4

The IEEE 802.15.4 group was formed with the express purpose of developing and defining a range of low-cost, low-power network layers (MAC + PHY), which could be used by a wide range of higher-layer protocols. The group does not attempt to specify what these are, leaving it to different standards groups to define according to their market application. By far the best known is ZigBee, but the underlying radio is also used by WirelessHART, RF4CE, MiWi,[6] ISA100.11a [7] and 6LoWPAN. 6LoWPAN is an interesting newcomer, as it uses standard IP to form embedded wireless networks.

The original version of 802.15.4 was released in 2003 and includes radios in three different bands, as shown in Table 6.2.

Later releases increased the data rates in the 868 MHz and 915 MHz spectrum and the most recent – 802.15.4a – adds modulation schemes that help the radio to be used for accurate location. It also brings in a UWB PHY that is specified in a number of bands, including the globally available spectrum from 6 GHz to 10 GHz. Readers interested in these should consult the IEEE standard.[8]

Table 6.1 *History of 802.15.4 and ZigBee releases*

802.15.4	
Version	Description
802.15.4–2003	The initial release, covering two different DSSS PHYs, one operating at 868 or 915 MHz and the other at 2.4 GHz. This version is used by all releases of the ZigBee specification.
802.15.4–2006	An update, which increased the data rate for the 868 or 915 MHz PHY. It also included four new modulation schemes – three in the lower band, and one at 2.4 GHz.
802.15.4a	A further two PHYs were defined; a UWB PHY and another 2.4 GHz PHY using chirp spread spectrum.

ZigBee	
Version	Description
ZigBee 2004	The original release, also known as ZigBee 1.0. Publicly released in June 2005. This is now deprecated.
ZigBee 2006	Released in September 2006. Introduced the concept of the cluster library.
ZigBee 2007	Released in October 2008. Contains two profile classes.
ZigBee PRO	ZigBee PRO is the new profile class within the 2007 release, and includes additional features for secure, robust deployments.
RF4CE version 1.0	A new standard released by the ZigBee Alliance in 2009 for remote-control devices.

Table 6.2 *Radios defined in 802.15.4–2003*

Frequency	Channels	Throughput (kbps)	Region
868 MHz	1	20	Europe
915 MHz	10	30	USA
2.4 GHz	16	250	Global

The lower-frequency bands of 868 MHz and 915 MHz have a significant practical advantage of offering greater range and consuming less power. However, they are not global. In particular, the limitation of a single channel within Europe makes these unattractive for manufacturers. As a result, almost all commercial products and chips use the 2.4 GHz option, so I will confine further discussion to that.

The RF specification for 2.4 GHz uses DSSS (direct sequence spread spectrum) for the radio. The principle behind DSSS is described inSection 5.3. The raw bit rate for the 802.15.4 radio is 2 Mbps. Rather than using this purely for achieving a high data transfer rate, the radio uses a chipping scheme, whereby 32 chips are used to represent every four bits of data. That has the effect of reducing the actual data throughput by a factor of eight, to 250 kbps. The advantage of this approach is that by using 32 chips to represent four bits of data, there's an effective gain of eight in the resistance to interference, making it easier to pull the data out of noise when the receive correlation is performed. That improves the receive sensitivity, and hence the link budget, and therefore increases the range.

For 802.15.4, the data throughput is normally irrelevant. Low-power radios are typically optimised to transfer occasional bits of data with low latency. Applications needing to transfer large amounts of data typically require power supplies or rechargeable batteries. That's not what low-power radio specifications are designed for.

Because networks formed from 802.15.4 radios normally contain nodes running off batteries, they limit their transmit power

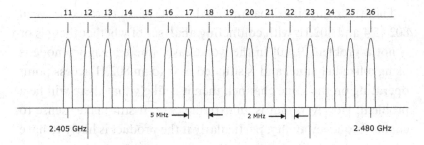

Figure 6.1 Spectrum usage for 802.15.4

to between −3 dBm and 0 dBm. The chipping scheme used helps compensate for this by providing better range due to the improved receive sensitivity. The basic receive sensitivity called for in 802.15.4 is −85 dBm for 2.4 GHz radios. Current chipsets offer real-life figures of between −90 dBm and −100 dBm.

Although range can be increased by increasing the transmit power, that is of little use in a bidirectional system if there are nodes still operating at 0 dBm. Although these will be able to hear incoming messages from other nodes transmitting at higher powers, the higher-power unit they reply to will probably be too far away to hear their weaker return messages.

DSSS radios operate at a fixed frequency. In the 2.4 GHz band, 802.15.4 specifies up to 16 channels, each 2 MHz wide, spaced 5 MHz apart (Fig. 6.1). Channel numbering starts at 11 for 2.405 GHz, up to 26 at 2.480 GHz. Lower channel numbers are allocated to the 868 MHz and 915 MHz bands.

Running at one frequency can make a radio susceptible to interference, particularly if it is using the same frequency as a nearby transmitter, such as an 802.11 access point operating on the same channel. 802.15.4 provides tools to allow a higher application layer to check for a clear frequency before choosing which one to use to commission a network. It also allows an application to implement the concept of frequency agility. This is where a network is able to monitor the state of the spectrum and, if it is suffering from interference, can migrate the whole network to a new, fixed frequency in a less congested segment of spectrum.

There is an ongoing debate about the issue of interference between 802.15.4 and 802.11, with conflicting analyses of whether there is or is not a problem.[9, 10] In the worst case, where a ZigBee node is using a fixed channel and is situated next to an 802.11 access point operating on the same channel, then it is likely that there will be a problem. However, that is an extreme case. It still makes sense to enable frequency agility, particularly if the product is likely to have a long operating life.

6.1.1 The MAC

The 802.15.4 MAC controls the flow of frames that pass through the radio and travel over the air. It is designed to accommodate many different network topologies and higher-layer stacks, offering security, guaranteed timeslots, beaconing services and node associations for forming a network. It can also provide validation services for frames. This means that the host only needs to be woken up for relevant frames. A consequence of the richness of the specification is that only a small portion of it tends to be used for any specific application. We will confine ourselves predominantly to the services relevant to ZigBee.

802.15.4 describes two types of network nodes – full function devices (FFD) and reduced-function devices (RFD). (These terms were initially used by ZigBee, but the more recent releases use more explicit descriptions of ZigBee Coordinator, ZigBee Router and ZigBee Endpoint.)

A reduced-function device is a simple end node – usually a switch or a sensor, or a combination of both. Reduced-function devices can only talk to FFDs as they contain no routing functionality. They're often referred to as child devices, which need parents to communicate. Their big advantage is that they are able to go to sleep for long periods, as they aren't needed to route messages around the network. Because they don't do much, they typically have a smaller stack and can be implemented at very low cost.

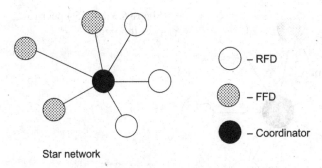

Star network

Figure 6.2 802.15.4 star topology

Full function devices do the heavy lifting in the network. In 802.15.4, FFDs are all capable of routing network data between nodes. They can also function as simple nodes, like an RFD. Depending on the volume, it may make sense for manufacturers to make only one version of node that can be set to act in either role.

A special form of FFD is the personal area network (PAN) coordinator. In addition to the functions of a standard FFD of routing message, the PAN coordinator is responsible for setting up and taking charge of the network.

6.1.2 Topologies

An 802.15.4 network can take two forms – a star network or a peer-to-peer network. A star network is shown in Fig. 6.2. Here a central PAN coordinator node is directly connected to a number of different nodes. Although all of the nodes can talk to the coordinator node, none of them can communicate with each other, even though some of them are FFDs themselves.

Figure 6.3 shows the same arrangement of nodes, but here the PAN coordinator has configured the network to be a peer-to-peer network. All of the previous direct connections between the nodes and the PAN coordinator remain, but they've been augmented with new connections (marked p2p) that allow the FFD nodes to talk directly to each other. Each of the three FFD nodes now has the ability to talk to other nodes, in addition to acting as a sensor node

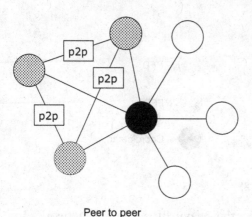

Figure 6.3 802.15.4 peer-to-peer topology

Peer to peer

itself. Note that there does not need to be any differences in the way the nodes are physically constructed. The only difference is that the PAN coordinator has allowed them to act as routers rather than simple FFDs.

Star networks can also be strung together with a backbone of FFDs to produce a cluster-tree network (Fig. 6.4). Each cluster-tree network requires one, and only one, PAN coordinator. Mesh networks can be implemented in a similar manner. But these require the formation and routing to be controlled by higher layers.

6.1.3 Framing

The 802.15.4 specification contains four basic frames: command frames, data frames, acknowledgements and beacons.

If a network based on 802.15.4 does not require QoS, it uses a standard CSMA/CA protocol with a random back-off. The CSMA is carrier sense multiple access; the CA is collision avoidance. It is a standard process for network access in situations where devices do not have negotiated timeslots for transmission. CSMA refers to the requirement that a node wanting to transmit starts by listening to the channel to see if it can detect any activity. If it cannot, then it is free to start its transmission. If it does detect activity, it sets a back-off timer and waits for that to expire. At that point it listens

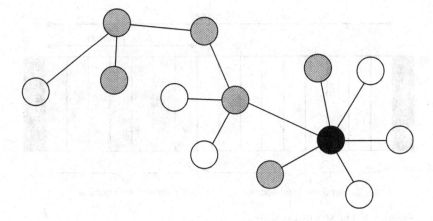

Cluster tree network

Figure 6.4 802.15.4 cluster-tree topology

again. If the channel is still busy, it increments its back-off time to a larger value and waits once again. The process is repeated with increasing back-off delays until the device is able to access the network. This technique is also known as clear channel assessment.

Depending on the back-off time and design of the radio, it may be able to enter a sleep state to save power during these periods.

The reason for adding a random exponential back-off timer is so that different nodes that want to transmit at the same time do not get into a situation where they all wake up after a back-off and try to transmit together, as that would result in them conflicting with each other for ever.

As well as the four basic frames, 802.15.4 also includes the concept of a superframe. The superframe is bounded by two beacon frames which are transmitted by the PAN coordinator.

The superframe is illustrated in Fig. 6.5. The beacons are transmitted without CSMA/CA, as the coordinator assumes that it has unconditional access to the wireless medium and won't suffer collisions. There are 16 timeslots between the beacons, during which any node can transmit, using a standard CSMA/CA scheme. Up to seven of the slots can be configured to be used for contention-free access, where nodes are assigned guaranteed timeslots. These are

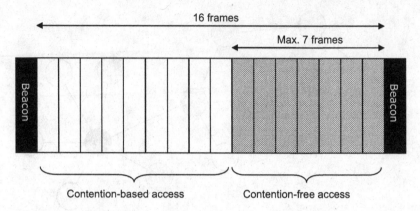

Figure 6.5 The 802.15.4 superframe

assigned by the PAN coordinator. These guaranteed slots can be used by stacks wanting to implement networks that require quality of service guarantees. Systems using superframes can elect to power the network down during unused portions of the superframe, including the coordinator. This can result in a power-efficient network. (Superframes are not utilised by ZigBee.)

6.1.4 802.15.4 security

ZigBee security is covered in Chapter 3. The 802.15.4 standard includes 128-bit AES. It provides protocols to use this at the baseband level, and also exposes the AES engine for use by higher layers.

The 802.15.4 standard is a large and complex one. For more details you can read the standard [8] or consult one of the many books covering it. My personal favourite for a practical overview of 802.15.4 and ZigBee is Drew Gislason's book.[11] It is one of the few aimed at designers.

6.2 ZigBee

ZigBee is far and away the best known of the networking standards that sit on top of the 802.15.4 MAC/PHY. It provides a mesh

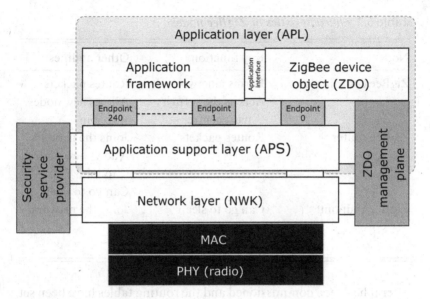

Figure 6.6 The ZigBee stack

networking capability, bringing redundancy and extended range to applications.

A good starting point is the architecture stack, which will be familiar if you've ever delved into the standard. It's shown in Fig. 6.6. The key components of it include the network layer (NWK), the application support layer (APS) and the application layer (APL) with its profiles. Alongside these, the ZigBee device object and security manager handle commissioning and security of the network. Rather than working through the individual layers, it is easier to look at how the network works and then see how they fit.

Before we start, it's worth revisiting the key features of a mesh network. ZigBee used to utilise the RFD and FFD terminology of 802.15.4, but in more recent releases has progressed to more descriptive and better-defined names. Every ZigBee node is one of three types – a ZigBee endpoint (ZED), a ZigBee router (ZR) or a ZigBee coordinator (ZC). There may be any number of routers and endpoints, but only one coordinator within any network. Unlike an 802.15.4 network, the ZigBee coordinator can leave the network

Table 6.3 *Characteristics of ZigBee nodes*

Node	Key functions	Other abilities
ZigBee coordinator	Forms a network	Routes packets
	Acts as the security trust centre	Allows new nodes to join
ZigBee router	Routes packets	Joins the network
		Allows new nodes to join
		Can go to sleep
ZigBee endpoint	Can go to sleep (allows battery operation)	Joins the network

after it has been commissioned and the routing tables have been set up in the routers. However, the network will then be unable to add any nodes or change its routing characteristics.

The three types of ZigBee node can be described by what they are able to do, which is listed in Table 6.3.

ZigBee endpoints are optimised to spend most of their life asleep. They are often referred to as child nodes, which need a parent – either a router or coordinator to talk to. Because of their low power, they are frequently battery-operated: a well designed ZigBee endpoint can run off an AAA battery for many years. ZigBee endpoints cannot route data to any other node – they can only talk to their parents. If those parents disappear or move out of range, then they have the ability to find and connect to a new parent.

In contrast, routers and the coordinator are generally always powered, ready to receive packets and either act on them or forward them to their eventual destination. It is possible for routers to sleep when the network is using beacons to enable timing.

The topology of the mesh is illustrated in Fig. 6.7. At first sight it looks similar to the peer-to-peer topology that we saw for 802.15.4; however, there are subtle but important differences.

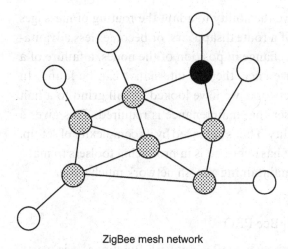

Figure 6.7 A ZigBee PRO mesh

ZigBee mesh network

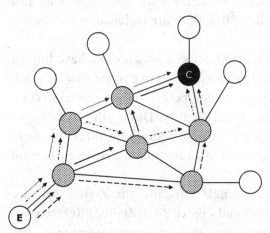

In ZigBee PRO, the send and return routes may be different

Figure 6.8 Multiple routes in a mesh

The biggest one is not visible on any diagram like this, but is implicit in the networking capability provided by the NWK layer, which is that messages may take multiple routes between the sending and receiving nodes. Figure 6.8 shows some of the possible routes that data could travel between the end node, E, and the coordinator, C.

Mesh networks have the ability to adapt the routing of messages dynamically, so that if a route disappears, or becomes less advantageous as a result of a change in position of the nodes, a failure of a routing node or interference, then an alternative can be found. In contrast, the other networks we have looked at will grind to a halt and at best alert a user that maintenance is required. This gives a mesh excellent reliability. That's balanced by its complexity of set-up. Much of what ZigBee has achieved is in producing toolsets to make commissioning and maintaining a mesh network much easier.

6.2.1 ZigBee and ZigBee PRO

Before talking about how a ZigBee network functions, it is worth explaining the differences between ZigBee 2007 (normally just called ZigBee) and ZigBee PRO. Both are included in the ZigBee 2007 release.

ZigBee is designed for small devices, which often have limited processing capability and memory, so it is important that the stack is small. However, this provides conflicts with the topological complexities of mesh networks and security. During the evolution of ZigBee it became clear that there were desirable features for a mesh network that were not implemented, but which were likely to push up the size of the stack.

Rather than produce a single standard, the ZigBee Alliance released two – ZigBee 2007 and ZigBee PRO. ZigBee PRO incorporates some major enhancements, which are not present in ZigBee 2007. The most important ones are:

- The maximum number of hops across the network is increased from 10 to 30. To achieve this, a new stochastic addressing scheme is employed.
- Multicast capability is supported, allowing a node to send a message to a predetermined set of destinations.
- Source routing is included. This is a technique that allows a coordinator or router device to request routers to find and temporarily remember the route for a new message. Because it

is a dynamic technique, it reduces the size of the routing tables needed in router nodes. This makes it practical to support larger networks without a massive memory overhead in each router.

- Asymmetric routes. This allows the acknowledgement to a message to be conveyed over a different route from the original message. This is particularly useful where there is asymmetry in the link budget between two nodes. Without it, it may be possible to transfer a message in one direction, but not the other.
- The total number of devices that can be supported in a network increases from 31 101 to 65 540, although this is probably only of academic interest.
- High security is added to the standard. This means that link keys can be supported at the application layer.

Within a network, support for either ZigBee 2007 (or earlier) or ZigBee PRO is indicated by the value of the stack profile, which is set to 0x02 for ZigBee PRO and 0x01 for all other releases.

Both ZigBee 2007 and ZigBee PRO contain significant improvements over earlier versions:, in particular, they now include support for frequency agility, where the entire network can move to another channel if interference is detected. I would strongly suggest that this should be considered as mandatory for any design.

There is not full backwards or forwards compatibility between ZigBee PRO and ZigBee 2007 because of the difference in routing techniques. ZigBee PRO routers cannot act as routers on a ZigBee 2007 network, but can work as ZigBee endpoints. Conversely, ZigBee 2007 routers can only exist as endpoints on a ZigBee PRO network. I see this as another reason for the industry to standardise on ZigBee PRO and provide full interoperability within the marketplace.

With two different versions targeting the same use cases, a designer needs to decide which to choose. Some companies involved with ZigBee recommend the use of ZigBee 2007 for devices with limited capability – often devices aimed at the consumer market – and ZigBee PRO for more professional products. I would argue that it makes sense to use ZigBee PRO in all designs, unless the ecosystem you are working with has already coalesced around ZigBee 2007. The enhanced security, along with the other enhancements of ZigBee PRO, means that it is a considerably more robust specification. The added cost should be more than offset by its reliability in the field. Some application profiles now mandate the use of high security and hence ZigBee PRO.

The general philosophy of both ZigBee 2007 and ZigBee PRO remains the same. For the rest of this discussion, the examples use ZigBee PRO, unless otherwise noted.

6.2.2 The ZigBee network

A mesh has very different properties from other networks. Most other networks will normally only support one unique conversation between devices at a time, albeit that this is often hidden by multiplexing links. In contrast, a mesh is better visualised as a cloud resource, where different pairs of nodes can be having concurrent conversations. Conceptually, it is a collection of many independent links sharing the same infrastructure. Although it is created and commissioned as a complete entity, once it starts operating, the individual links are often independent of each other. What the mesh provides is a redundant shared connection medium.

The basic concept of how a mesh network operates is simple. It is about addresses and routing. To make that concept a reality, mesh standards need to define reliable methods, which determine how data and commands get from one node to another, when the journey involves multiple hops.

Trying to explain this is far from straightforward. One way to approach it is by understanding the different levels of addressing and the services that each of them provide. I'll start with that and then describe how these link in with the stack architecture shown in Fig. 6.6.

Each ZigBee network consists of a number of nodes that operate together on one radio frequency or channel. This network is known as a PAN (personal area network) and is assigned a PAN

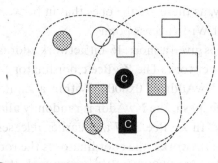

Figure 6.9 Overlapping PANs

○ PAN A – Channel x

▢ PAN B – Channel y

ID. ZigBee allows up to 16k different PAN IDs, ranging from 0x0000 to 0x3fff. The PAN ID is normally a random number, which is generated when the network is first set up. Private PAN IDs are allowed for proprietary applications, where a manufacturer can specify an extended PAN ID (EPID), which is a 64-bit number.

As ZigBee supports 16 channels in the 2.4 GHz band, there can be several networks in the same physical space, each operating on a different channel. In previous versions of ZigBee, prior to the introduction of frequency agility, it was possible to use the same PAN ID for networks operating on different fixed channels. That is no longer allowed. There is nothing to prevent multiple networks existing on the same channel (Fig. 6.9), but it obviously increases the risk of interference and is best avoided.

The channel for the network is assigned at the point of network creation. The coordinator node forming the network first listens to each channel to see what activity is present (which may not be ZigBee activity). It then sends an active probe out to any ZigBee networks present on each channel to determine how many are operating on each one. Using this information, it chooses a channel, giving priority first to the channel with the fewest ZigBee networks, followed by the quietest channel. Some application profiles may impose different criteria. It is not unusual to choose channels

15, 20, 25 or 26 preferentially, as these are the ones that fit between the default channels for most Wi-Fi access points.

Each node within the PAN is given a short, 16-bit network address (NwAddr) when it joins the network. The ZigBee c,oordinator for the network always has an NwAddr of 0x0000. Within a ZigBee PRO network, which is a true mesh, the NwAddr is randomly allocated as it joins the network. In ZigBee 2007 and earlier releases, which implement cluster-tree networks, the coordinator is the root of the tree with an NwAddr of 0x0000. ZigBee routers are then assigned addresses starting at 0x0001 and ZigBee endpoints have addresses from 0x0796.

The NwAddr allows devices to be addressed within a PAN without the overhead of the full 64-bit MAC address. However, all devices need to keep tables that cross-reference the MAC address to the NwAddr. That allows the network to cope with the eventuality that a device becomes disconnected from its parent and needs to rejoin the network. If this happens, it will be allocated a new, different NwAddr. It then needs to alert all other nodes in the network to this change, so that they can update their tables.

Sets of nodes can be allocated to groups. This allows broadcast messages to be sent out and acted upon by more than one node. A typical application is a switch that turns on a number of different lights. That conveniently brings us to the two main messaging modes – buroadcast and unicast.

6.2.2.1 Broadcast messages

Broadcasts are used to convey the same message to a number of different nodes. They are received by all nodes, but will only be acted upon by those nodes for which they contain a relevant message. Broadcasts can also be configured so that they are only sent to nodes that are awake (this means that they don't need to be buffered by parent nodes whose children are asleep), or they can be limited to being sent only to routers (plus the coordinator). Broadcasts are a powerful way of sending data to more than one device at the same time, but they come with a few provisos.

The first limitation is that broadcasts are not acknowledged. That is a standard feature of any broadcast transmission – the network would be overloaded if every device receiving it tried to respond. The second is that they can clog up a network if overused. To help cope with that, the maximum number of hops for a broadcast can be set, to limit how far it propagates through the network.

6.2.2.2 Unicast messages

In contrast, unicasts are messages that are directed to a specific, unique NwAddr and typically a specific endpoint on that device. They allow acknowledgements from the destination device.

6.2.2.3 Multicast messages

ZigBee PRO also supports multicasts, which are a type of broadcast that is only sent to members of a specified group. They use a more efficient method of retransmission by routers, which limits their routing to routes that will reach group members. Where a group is dispersed throughout a network a limit to the number of intermediate nodes can be specified.

Broadcasts are inherently slow. As a rule of thumb, it takes around 10 ms for a unicast message to propagate between two nodes in a ZigBee mesh. As soon as a router receives the message it looks up its next destination and sends it on. So a unicast message sent across 10 nodes will arrive in around 100 ms. In contrast, when a router receives a broadcast it checks to see whether it has already received it and how many hops it has already traversed. If it determines that it needs to be forwarded, then it adds it to its broadcast transaction table (BTT) and then rebroadcasts it. This is a much slower process, which means that using broadcast to send a message across ten hops could take around ten times as long, or close to one second. If a router receives too many broadcasts, it may exceed the capacity of its BTT, so broadcasts should be used sparingly.

Broadcasts are used for some of the fundamental management function of a ZigBee network, in particular determining routes. ZigBee does this by utilising a technique based on the public advanced

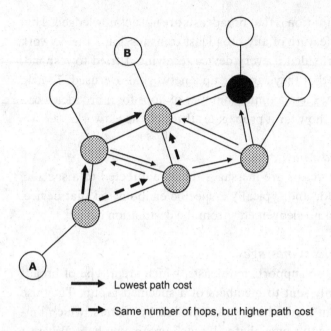

Lowest path cost

Same number of hops, but higher path cost

Figure 6.10 Route discovery

ad hoc on-demand distance vector-routing algorithm (AODV).[12]
When the network is formed, it has no knowledge about the relative
locations of different nodes. So when a node wishes to communicate
with another node, it needs to use a route-discovery process.

6.2.2.4 Route discovery
Route discovery starts when the sending node sends out a broadcast message to its intended destination.

In Fig. 6.10, node A wishes to find a path to node B. At each
routing node where the message is received, the router calculates
a 'path cost', which is a value for the quality of the previous link.
It adds this to the packet and retransmits it after a random delay.
Each router also keeps details of the route in its temporary route-
discovery table. As the message traverses its route, each router
adds to the path cost, based on the quality of the link. Node B will
eventually receive these packets, which have arrived via different
routes, and look to see which has the lowest cumulative path cost.

The lowest path cost may often not correspond with the shortest number of hops. Three short hops with good link budgets may be better than two long hops with poor link budgets, as the latter will involve multiple retries.

Node B then sends details of the route with the lowest path cost back to node A using a unicast message along that route. The routers along this path update their routing table so that they can deliver packets to this node on future occasions. Routers that aren't on this route will discard the information from their temporary route-discovery tables and won't update their main routing table. Note that ZigBee endpoints are not involved in this process. Their parents do the work for them.

In all versions of ZigBee, except for ZigBee PRO, routing is symmetric. In other words, packets travelling in either direction between two nodes follow the same route. In ZigBee PRO a different return route is allowed. This can help where the link budget between nodes is asymmetric, which may be the case if they have different transmit outputs or receive sensitivities. As a general rule, it is good practice to design all of the nodes in a ZigBee network to have the same transmit output and receive sensitivity, otherwise you may find nodes that can only communicate in one direction. Once discovered, the same route is maintained until it fails, or an application forces a new route discovery.

Because the broadcasts needed for route discovery are fairly resource intensive, it is good practice to commission devices one at a time to prevent the network becoming overloaded. They also need resource to keep themselves updated if the network regularly allows other devices to join. Keeping routing tables updated is an important consideration in consumer products. It needs to be automatic and foolproof.

6.2.3 ZigBee profiles and applications

Mesh networks usually contain a mix of different devices, which want to share information with each other, or possibly transmit it via a gateway. Typically, many of these devices are battery-powered, with limited hardware capability, so it is necessary to find an efficient way of defining how they carry out their functions and with whom they communicate. ZigBee manages this by organising general application profiles which cover product ecosystems. These in turn use the ZigBee cluster library to define individual, common actions.

Within each ZigBee device, endpoints define the applications that it is capable of performing (not to be confused with a ZigBee endpoint device). Each device can have up to 240 endpoints, numbered from 1 to 240, and each of these contains information about the application it can perform. It is useful to think of an endpoint as a sub-address within the node, as it is the ultimate destination where commands are generated or consumed. Endpoint 0 within each device is the endpoint for the ZigBee device object (ZDO), which is responsible for joining the device to the network and specifying what sort of ZigBee device it is (coordinator, router or endpoint). The ZDO does this by using the special ZigBee device profile (ZDP).

Going back to the device's endpoints, each of these contains three key items:

- A profile ID, which defines which of the ZigBee application profiles it supports,
- A cluster ID, which defines which cluster from the ZigBee cluster library it uses, and
- A device ID, which defines the physical entity that this application represents, such as a light switch or a thermostat.

The core concept is the ZigBee cluster library or ZCL. The ZCL is a list of common actions that are used across many applications. These include actions like on–off, time, temperature, etc. Each cluster has a universally unique identifier (UUID). All clusters with that UUID will perform that specific function in the same manner, regardless of which public or private profile they are used in.

Each cluster contains a predefined number of attributes, each again having its own ID, although this ID is not global like the

Table 6.4 *Examples of ZigBee clusters*

Cluster ID	0x0000		Basic cluster
	Attribute ID	0x0000	ZCL version
		0x0001	Application version
		0x0002	Stack version
		0x0003	HW version
		0x0004	Manufacturer name
		Etc.	
		0x0010	Location
		Etc.	

Cluster ID	0x0002		Temperature
	Attribute ID	0x0000	Current temperature
		0x0001	Minimum temperature
		0x0002	Maximum temperature
		Etc.	
		0x0010	Temperature alarm mask
		Etc.	

cluster ID, but just has meaning within that specific cluster. The examples given in Table 6.4 indicate the structure:

The attribute number sent over the air defines exactly what the attribute does, as defined within the ZCL specification. For clarity in the definitions, attributes are divided into sets of related functions. Each set is indicated by the first 12 bits of the attribute and is called an attribute set. In the examples in Table 6.4, the first two attribute sets for the basic cluster are the basic device information, starting at 0x0000 and the basic device settings, starting at 0x0010. For the device temperature configuration example, they are the device temperature information, starting at 0x0000 and the device temperature settings, starting at 0x0010.

ZigBee profiles define devices within an ecosystem, specifying which of these clusters each device must support. All devices supporting that profile must understand and implement these in order to ensure interoperability. The device ID that we came across in this example provides an identification of the endpoint, which can be used in a commissioning tool, allowing such an application to display end-user information to guide the process.

With the ZigBee cluster library and the knowledge of addressing and routing, we can start to put the ZigBee model together. The application support layer (APS), which sits on top of the networking layer (NWK), is the layer that enables applications to talk to each other.

The APS looks at each packet passed up to it by the NWK layer and filters out those that are not appropriate, passing on valid packets to the relevant endpoint or group of endpoints. It takes responsibility for generating ACKs for multi-hop unicast messages. It also runs the management tasks that establish the application links between devices:

- It maintains the **local binding table**. This provides the links between endpoints on two devices. These are unidirectional and stored at the sending node for each link. Each entry contains the source endpoint, the destination NwAddr and endpoint and its cluster ID.
- Where appropriate, it maintains the **local groups table**, listing which endpoints within the device belong to a group.
- It keeps an **address map** of target network addresses and their associated MAC addresses. When a device leaves the network and rejoins with a new NwAddr, this table gets updated when the device broadcasts its new NwAddr, allowing communication between endpoints to continue uninterrupted.

The process of binding is not mandatory. Transactions can be sent directly, but deploying and maintaining a network is much simpler if bindings are used.

To summarise, Fig. 6.11 provides a highly simplified representation of how the different layers of the ZigBee architecture build up the information level to the point that a light switch knows the difference between a light bulb and a socket within a home automation network and how to control the correct bulb.

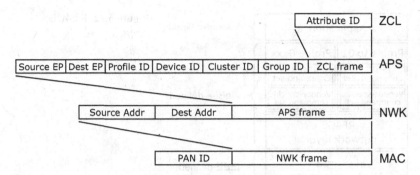

Figure 6.11 Simplified framing hierarchy for ZigBee

It is worth emphasising that ZigBee's concept of profiles is much wider than for other wireless standards. Whereas other standards generally define profiles that apply to a specific application, such as a Bluetooth headset or printer, the ZigBee Alliance writes profiles covering complete ecosystems of devices.

The two public profiles that have been released at the time of writing are the home automation profile, which covers lighting, heating, ventilation and air conditioning (HVAC) appliances around the home and the smart energy profile, which covers smart meters, displays and appliances. ZigBee only has a few profiles, but these are made much wider-ranging in their applicability through the use of the ZigBee cluster library that is used within them.

6.3 ZigBee RF4CE

ZigBee RF4CE is a standard produced by the ZigBee Alliance for remote control devices, predominantly in the home audio and video goods market. It aims to provide a replacement for traditional infrared remote controllers, offering a number of advantages:

- A bidirectional data transfer, so that remote controls can receive information from their target device,
- Increased security,
- The ability for a remote control to operate with multiple target devices.

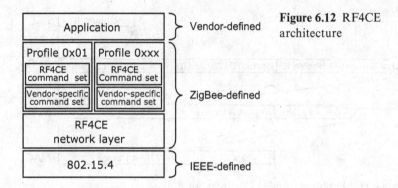

Figure 6.12 RF4CE architecture

Like ZigBee, the standard is based on 802.15.4. Although it takes many of the features of the ZigBee standard, RF4CE is a much simpler stack, enabling it to be used in very low-cost remote controls. The overall architecture is shown in Fig. 6.12.

To reduce complexity, all of the work is done by the network layer, without a ZDO or APS. Although the RF4CE standard is defined by the ZigBee Alliance, this means that RF4CE devices are not compatible with ZigBee mesh devices, although they can coexist in the same wireless space.

More information on RF4CE can be found in the specification and white papers released by the ZigBee Alliance.[3, 13]

6.4 6LoWPAN

One other 802.15.4 network attracting attention is 6LoWPAN. The name is an acronym for 'IPv6 over low-power wireless personal area networks'. The concept behind 6LoWPAN is simple – it's to bring IP directly down to small, low-cost sensor devices. Acknowledging that there are not enough of the current IP-format addresses within the world to extend to the 'Internet of things', 6LoWPAN starts from the premise of IPv6, with the aim of giving an address to every device.

That creates a problem; if a set of 40-byte IPv6 addresses were to be placed in a 127-byte 802.15.4 frame, there would be precious little room for any payload. To get around this, the LoWPAN networking

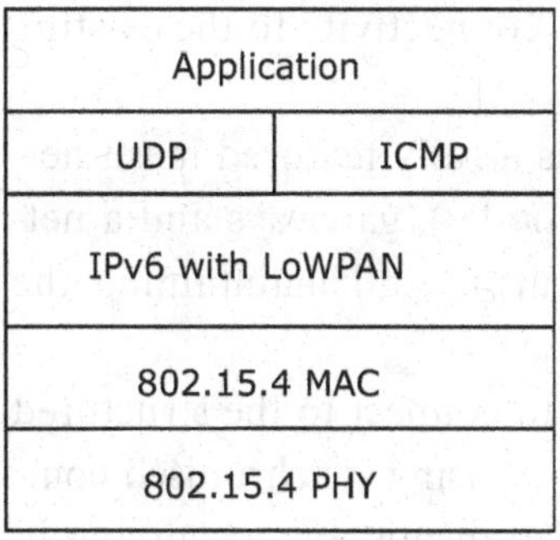

Application	
UDP	ICMP
IPv6 with LoWPAN	
802.15.4 MAC	
802.15.4 PHY	

Figure 6.13 6LoWPAN architecture

layer uses stateless address compression to reduce the address to a handful of bytes.

The stack (Fig. 6.13) specifies UDP and not TCP as a transport, further limiting any unnecessary clutter in the packets. The result is a very compact stack, which is significantly smaller than a ZigBee mesh stack.

6LoWPAN is being developed by the Internet Engineering Task Force (IETF) as an open standard. It is still in its early days, but is attracting considerable interest, as it promises to make it very simple to extend existing IP networks to individual sensor nodes. It is creating a lot of interest within the smart energy and smart grid movement, where IP connectivity is seen by NIST (the National Institute of Standards and Technology) [14] as an important advantage. It is not coincidental that the next release of ZigBee's Smart Energy profile is likely to include IP connectivity.

The 6LoWPAN specifications can be downloaded from the IETF site. The working group is generating two documents, which define the application and the adaptation layer.[15, 16]

6.5 WirelessHART

Another 802.15.4 standard gaining traction is WirelessHART.[4] This is a more specialised standard, which is used in factory and process control, where it adds wireless connectivity to the existing wired HART standard.

Like ZigBee PRO, WirelessHART is a fully featured mesh network containing field devices (sensor nodes), gateways and a network manager responsible for configuring and maintaining the network.

Because WirelessHART is designed to connect to the structured HART protocol, it takes the approach of using synchronised communications between devices, with all transmissions occurring in pre-scheduled timeslots. That also allows it to implement channel hopping, with automatic, coordinated, hopping between channels to increase its immunity to interference. As a result, it can define QoS for transmissions, which is important for process control applications.

Although WirelessHART is aimed at a specialised market sector, the number of devices deployed using the wired HART protocol means that it represents a significant volume of the 802.15.4 market. A number of silicon suppliers believe that by 2012 it may be responsible for up to one third of all 802.15.4 shipments.

6.6 References

[1] ZigBee Alliance, www.zigbee.org.
[2] IEEE WPAN 802.15.4 Task Group 4, www.ieee802.org/15/pub/ TG4.html.
[3] ZigBee Alliance, Zigbee RF4CE specification. www.zigbee.org/ ZigBeeRF4CESpeciification/tabid/464/Default.aspx.
[4] WirelessHART, www.hartcomm.org/.
[5] 6LoWPAN Working Group, www.ietf.org/dyn/wg/ charter/6lowpan-charter.html.
[6] David Flowers and Yifeng Yang, MiWi wireless networking protocol stack. www.microchip.com/stellent/ idcplg?IdcService=SS_GET_PAGE&nodeId=1824&appnote=en52 0606.
[7] ISA, ISA100.11a, Release 1: an update on the first wireless standard emerging from the industry for the industry. www.isa. org/source/ISA100.11a_Release1_Status.ppt#349.

[8] IEEE Standards Association, IEEE 802.15 wireless personal area networks. http://standards.ieee.org/getieee802/802.15.html.

[9] Z-Wave Alliance, WLAN interference and IEEE 802.15.4 (2006) www.zen-sys.com/modules/iaCM-DocMan/?docId=84&mode=CUR.

[10] ZigBee Alliance, ZigBee – WiFi coexistence (2008) www.zigbee.org/imwp/idms/popups/pop_download.asp?contentID=13184.

[11] Drew Gislason, *Zigbee Wireless Networking* (Newnes, 2007).

[12] C. Perkins, E. Belding-Royer and S. Das, Ad hoc on-demand distance-vector (AODV) routing (2003) http://tools.ietf.org/html/rfc3561.

[13] ZigBee Alliance, Understanding RF4CE (2009) www.zigbee.org/imwp/idms/popups/pop_download.asp?contentID=16212.

[14] Report to NIST on the smart grid interoperability standards roadmap: priority action plans – illustrative versions (2009) www.nist.gov/smartgrid/PAP_Combined_WorkshopFinalV1_0a_20090730.pdf.

[15] *RFC4919 – IPv6 over Low-Power Wireless Personal Area Networks (6LoWPANs): Overview, Assumptions, Problem Statement, and Goals.*

[16] *RFC4944 – Transmission of IPv6 Packets over IEEE 802.15.4 Networks.*

7 Bluetooth low energy (formerly Wibree)

Bluetooth low energy is the latest short-range wireless specification to appear on the market, having been ratified at the end of 2009. Although written by the Bluetooth Special Interest Group, it is a fundamentally different radio standard from the one covered in Chapter 5, both in terms of how it works and the applications it will enable. Hence it merits its own chapter.

By itself, Bluetooth low energy is incompatible with a standard Bluetooth chip – it is a completely new radio and protocol stack. Some of the applications it enables, such as allowing sports equipment to talk to watches, will use Bluetooth low energy chips for both ends of the link, neither of which will be able to talk to existing Bluetooth chips. In these end-to-end applications, it is not dissimilar to other low-power proprietary standards, such as ANT. [1] However, where it differs, and what gives it its power, is that the standard allows dual-mode chips to be designed, which support multiplexed Bluetooth and Bluetooth low energy connections. These will replace the Bluetooth chips in today's mobile phones and PCs, providing an infrastructure of billions of devices that can communicate with existing Bluetooth peripherals, as well as the new generation of dedicated Bluetooth low energy products. It gives Bluetooth low energy the 'free ride' that will lead to economies of scale for chip vendors and a vibrant ecosystem of devices for products to connect to.

Bluetooth low energy has a long history. The original incarnation of the standard was put forward by Nokia as one of the alternative proposals in the early stages of the 802.15.4 standard development. [2] It was not selected at the time, but continued to be developed as a low-power radio under the proprietary name of Blulite. In Octoer 2006, this was publicly announced as the Wibree standard, with support from a number of silicon vendors and device manufacturers. The following summer, ownership and development of the standard was passed from Nokia to the Bluetooth SIG, when it was renamed Bluetooth Ultra-low Power. It was renamed again in 2008 to Bluetooth low energy and finally released in December 2009.[3]

The reason for its existence was the desire of mobile phone manufacturers to extend the applications on their handsets to include personal sensors such as fitness devices, sports equipment, watches and ID tags. All of these have a requirement to operate off a coin cell for many months or years. Although they are already low powered, traditional Bluetooth products cannot achieve this extreme level of battery life because of Bluetooth's fast hopping, connection-oriented behaviour and relatively complex connection procedures. Nor was it possible to modify the existing standard to meet these requirements without breaking it. What was needed was a new standard that could coexist with Bluetooth, but which moved from a connection-oriented protocol to one where devices could sleep for most of their lives.

There was another practical constraint that shaped Bluetooth low energy. Phone manufacturers did not want to include another radio within their handsets because of cost and the space requirements of another antenna. Any new radio had to be able to lead a symbiotic existence with an existing radio in the handset, introducing no new components. Bluetooth low energy solved this problem by adopting a strategy where the new radio could share Bluetooth's radio structure within a dual-mode chip, which supports both standards. The low-power devices that connect to these use a single mode chip that only incorporates the Bluetooth low energy features. This allows very low-cost products, targeted at ultra-low power applications. As a result, phone manufacturers can add a new radio standard at no additional cost, whilst opening up a market for a wide range of low-power accessories for their handsets. I'll look at the detail of this dual-mode architecture after I've been through the basics of Bluetooth low energy.

7.1 Basic tenets

Bluetooth low energy is one of the few wireless standards to have had the luxury of starting largely with a clean sheet of paper. Although it had the constraint that it needed to be able to reside within a dual-mode Bluetooth chip, it had the opportunity of making independent decisions to deliver several key benefits. These explain much of the structure and development of the standard, including the manner in which low-power consumption is achieved.

7.1.1 Small packet size

Bluetooth low energy packets sent over the air are very small, ranging from 10 octets to a maximum of 47 octets. That includes data and commands. Such small packets are optimised for very concise 'chunks' of information, typically a single measurement or control action. Using packets this small means that there is a minimal amount of control information included, which limits them to performing fewer different tasks than most other standards. This simplicity means there is only one fundamental format for all of the different packets used by Bluetooth low energy.

A consequence of the small packet size is that Bluetooth low energy is not efficient at transferring large amounts of data, whether that is in the form of files, or repeated pieces of information. It is designed for intermittent events. If larger amounts of data need to be transferred, other wireless standards are likely to be more efficient.

7.1.2 Autonomous controller

A key technique in reducing power consumption is to allow as much of the device as possible to stay asleep. In Bluetooth low energy, the decision was taken to make the controller (the radio and MAC layers) as autonomous as possible, so that it only needs to wake up the higher host layers when absolutely necessary. This means that most of the device circuitry can remain in a deep-sleep mode for the majority of the time. The controller contains a connection whitelist and filtering, so that it is capable of discarding packets or duplicate messages without any intervention from the host controller.

7.1.3 Duty cycle and latency

As we have seen in Chapter 2, minimising power consumption is all about staying asleep for as much of the time as possible. Bluetooth low energy has been optimised to allow devices to do this, and perform a minimal number of transactions when they wake. The connection set-up and data transfer phase for a Bluetooth low energy connection can be as short as 3 ms.

7.1.4 Asymmetry

Bluetooth low energy is largely asymmetric in terms of the capabilities of the two ends of the link. The specification makes the assumption that in most cases, the sensor device will have very limited resources, in terms of power supply, processing capability and memory, whilst the receiving device is likely to be considerably better equipped. There is no need for devices to be able to support both a slave and a master role. This allows very simple low-power devices to be constructed.

7.1.5 Range

As many of the applications of Bluetooth low energy involve sensors that are located around a normal home, the range of the wireless connection needs to cover this. It also has to cope with interference from other devices operating within the same band, as well as the issues of multipath fading.

7.1.6 Ease of use

Unusually for a wireless standard, ease of use for designers was taken into account. Bluetooth low energy has been developed with the view that it should be straightforward for designers to use it in a wide variety of different applications. It was envisaged that many devices would connect via mobile phones to an Internet site, so support has been included for generic gateways. The philosophy of ease of use has also been extended to a simpler, cost-effective, qualification and verification scheme.

7.2 RF

The radio specification of Bluetooth low energy is necessarily constrained by the requirement that it can be implemented using the same RF chain already present in a standard Bluetooth chip. Despite that, it permits single-mode chips to be made with remarkably low sleep current, which can achieve an acceptable range with limited transmit power. This latter restriction is to allow chips to be powered by coin cells, where the peak current must be limited to less than 15 mA to prevent degradation of the cell.

The radio supports 40 channels in the 2.4 GHz band, each 2 MHz wide. These use GFSK (Gaussian frequency-shift keying) with a modulation index of 0.5 and bit-time product of 0.5. This is relaxed, compared with Bluetooth BR/EDR, and helps to increase the operating range. The overall RF specification is similar to that used of other ultra-low-power proprietary radios.

7.3 Topology

The topology of Bluetooth low energy is simple: it only supports piconets. The first release covers point-to-point connections, although it allows the implementation of a generic gateway profile on a central device, which can be used to send information to a remote service. Future releases of the specification are likely to extend the topology with switch and relay functionality, to enable the construction of star networks, using Bluetooth low energy, or another network transport, to form an extended backbone.

In the majority of envisaged use cases, the topology is based on devices that push information by broadcasting, advertising or planned notifications. Where connections are established, devices normally sleep for most of their lives, waking at pre-agreed times to exchange information. The simplest devices, which only broadcast or receive information, may consist of just a transmitter or a receiver, without the need for both.

7.3.1 Profile roles

Bluetooth low energy has introduced the concept of profile roles to describe the basic capabilities that devices have. The term is a little confusing, as it has nothing to do with application profiles, but describes the functionality a device has when communicating with other Bluetooth low energy devices.

Profile roles can be split into two categories. First, there are unidirectional devices where data are either sent or received with no acknowledgement. Second, there are bidirectional devices, where a conversation can be held between the two devices.

7.3.2 Unidirectional devices

The simplest types of device support either a broadcaster (transmitter) or an observer (receiver) profile role. Broadcast devices (Fig. 7.1) need only contain a transmitter: they send advertising packets, which contain data, and which can be heard by any receiving device. These data are identified using a universally unique ID (UUID), which

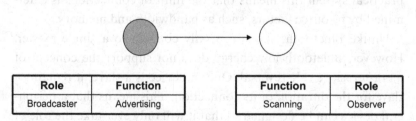

Role	Function		Function	Role
Broadcaster	Advertising		Scanning	Observer

Figure 7.1 Low energy broadcasters and advertisers

defines its format and type. The packets are transmitted according to the application running on the broadcast device. This may be a timed event, such as a regular transmission of temperature, or event-driven, as in the case of a footfall from a running shoe.

The equivalent receive-only device is called an observer, which does not need to contain a transmitter. Observers listen for broadcast messages by scanning for advertising packets. They can filter these according to a pre-installed application and then display or otherwise process the received data. They cannot send any acknowledgements or instructions back to the transmitting device.

Broadcast and observer devices offer the ability to produce the lowest-cost devices with the longest battery lives, but because they are unidirectional, they need to be preprogrammed, or controlled by an external application. They target a limited number of high-volume applications, where both cost and power are critical.

7.3.3 Bidirectional devices

Most Bluetooth low energy devices contain both a receiver and a transmitter, allowing them to negotiate with each other. These devices are described as supporting peripheral or central profile roles. Almost all Bluetooth low energy products fall into this category. After a connection is made, a peripheral device will take the role of a slave, whilst a central device will become the master.

A master can be connected to many slaves. In Bluetooth low energy, the number of connections is limited by the access address, which equates to a theoretical number of around two billion. In any practical system this means that the limit of connections is determined by resource factors, such as bandwidth and memory.

Unlike Bluetooth, slaves can only connect to a single master. However, Bluetooth low energy does not support the concept of a master–slave role reversal. Once a device is a slave, it remains a slave for the duration of its connection. This means that a peripheral device can be designed so that it will only ever take the role of a slave. The consequence of this is that it allows a major asymmetry

Role	Function		Function	Role
Peripheral	Advertising		Scanning	Central
Peripheral	Advertising		Initiating	Central
Peripheral	Connected (slave)		Connected (master)	Central
			Initiating *	Central
			Scanning *	Central

* A master can be scanning or initiating with other peripherals whilst in a connection

Figure 7.2 Low energy peripheral and central profile roles

in the complexity of peripheral and central devices. Slave devices can be much simpler (and cheaper) to produce.

There is no reason why a device cannot move between different profile roles. For example, it could start life in a peripheral profile role, allowing it to receive configuration instructions from a master device. Then, on completion of its configuration, it can turn off its receiver (possibly permanently), and transition to a broadcast profile role. However, in this state it would lose any connection with the master and be reduced to broadcasting data promiscuously. It is possible for a device to support multiple roles at the same time, including being both a master and a slave.

7.4 Advertising and data channels

At this point it is important to understand how Bluetooth low energy uses the spectrum. To provide robustness in the licensed 2.4 GHz band, Bluetooth low energy uses an adaptive frequency-hopping scheme. The standard splits the spectrum into a total of 40 channels: 37 of these are used for data transmission; with the other three used for fixed advertising channels. These three channels are used to broadcast data (advertising mode), to discover

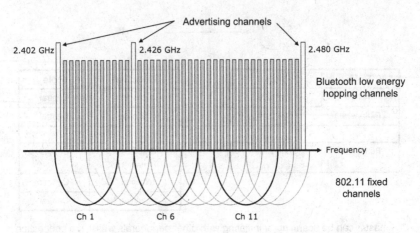

Figure 7.3 Bluetooth low energy frequency channels

other devices (scanning mode) and to make connections (initiating mode).

The three advertising channels are located at the two ends of the band – 2.402 GHz and 2.480 GHz – with the third at 2.426 GHz. These frequencies are chosen to avoid the areas of the spectrum used by 802.11 (2.426 GHz is located just at the bottom of 802.11's Channel 6). Figure 7.3 shows how they are arranged to avoid interference with the three most commonly used 802.11 channels – channels 1, 6 and 11.

Advertising is the basis for how Bluetooth low energy operates. Broadcaster and peripheral devices spend most of their time asleep and, on waking, use these channels to control much of their data transmission and connectivity.

Data channels are only used within connections, either for service discovery or directed data transfer. Data channels are like cables; they carry data between two selected devices rather than broadcasting it promiscuously. The data channels employ adaptive frequency hopping, moving to a new frequency at each new data event. This makes them more robust. Broadcaster and observer devices operate without ever using a data channel. (To do so they would need to take on peripheral and central pofile roles for the purpose of configuration.)

Preamble	Access address	Payload	CRC
1 octet	4 octets	2–39 octets	3 octets

Figure 7.4 Bluetooth low energy packet structure

Advertising occurs during advertising events. During an advertising event, a device, acting in the broadcaster or peripheral role, sends out an advertising packet, generally repeated on each of the three advertising channels. These use fundamental link layer packets, which may contain additional data, but which also signify the status of the device sending them. They inform any listening device whether the advertiser is discoverable and connectable, and whether more information can be requested without the need to make a connection.

A central device in receipt of these packets may be able to respond with two other fundamental link layer packets, either requesting more information, or asking to initiate a connection.

Although these packet types are hidden deep in the link layer, they form the fundamental basis of Bluetooth low energy connections, so we'll look at them in a little more detail.

7.4.1 Advertising packets

The packet design (Fig. 7.4) in Bluetooth low energy is very simple, with a single packet type for all transactions. Each packet is between 10 and 47 octets long and contains:

- A preamble, either 01010101 or 10101010,
- A 32-bit access address,
- The payload, which can range from 2 to 39 octets,
- A 24-bit CRC, which is calculated over the payload.

For this discussion, the important part of the advertising packet is the payload, which is aptly called the advertising channel PDU (protocol data unit). This consists of a two-byte header and an advertising payload of between 0 and 37 octets. The header indicates the length of the payload, whether the addresses within the packet are public or private and the type of advertising packet, which can be one of four types.

- ADV_NONCONN_IND (non-connectable advertisement):
  These are used to send advertisements containing data, but do not elicit any response from a device receiving them.
- ADV_DISCOVER_IND (discoverable advertisement):
  These contain data and may indicate that more data can be obtained if requested.
- ADV_IND (advertisement):
  These advertisements may indicate that more data are available and also tell the receiver that the device is available to make a data connection.
- ADV_DIRECT_IND (directed advertisement):
  These are special advertisements sent to a known master (central device) to request the re-establishment of a connection. They are sent with very low latency.

7.4.2 Response packets

A central device listening for advertising packets has a number of options when it receives one. It can choose to ignore the packet, it can accept the data within it and pass it up to the host stack, or, if the advertising packet allows, it can respond to the advertiser.

7.4.2.1 Scan response packets

If the cental device wants to receive more data, it responds with a scan request.

- SCAN_REQ:
  A request for further information to be sent, where the advertisement contained information saying this is available.

When the advertiser receives a SCAN_REQ, it immediately sends the additional data using a scan response packet.

- SCAN_RSP:
A packet that contains the additional advertisement data.

Scan requests are used where an advertiser indicates that it has more information than will fit into the 31-byte payload of an advertising packet. (As each payload has been fitted into a higher-level packet, we've used up 16 of the original 47 bytes for control and management tasks, so there is only a maximum of 31 bytes left for data. It is a similar principle to that of Russian stacking dolls.)

When scan responses are used to send additional data, it is good practice to arrange the payload so that regularly changing data are included in the initial advertisement, and static data are placed in the scan response. An example might be a thermostat, where the actual temperature would be placed in the original advertising packet and the set-point or thermostat's location in the scan response. A central device could understand that this information rarely changes and only ask for a scan response occasionally. If the data were presented the other way around, the master would have to use a scan request for every advertising packet, to obtain the current temperature, which would double the number of packets the thermostat would need to send. Attention to fine detail like this in a design is important if the maximum battery life is to be attained.

7.4.2.2 Initiating packets

If the central device is in the initiating state and receives an advertising packet, which informs it that the advertiser can make a connection, then it can initiate this by sending a connection request.

- CONNECT_REQ
A request to make a connection where an advertisement has stated that this is allowed.

When the advertiser receives this packet, it can choose to ignore it. That may happen because the responding initiator's address is not in its whitelist. If it accepts it, it termintes its current advertising event and starts the connection procedure.

The connection request from the initiator contains all of the link layer information that the peripheral requires to make a connection. This includes the access address for the central device; a value to initialise the CRC register; the hopping-sequence information, including a list of channels to use, and the connection interval times; plus a value for the master's sleep clock timer accuracy. The interval time is critical, as it determines how often the peripheral device wakes, and hence is a major factor in determining its battery life. Its value should be carefully considered.

An important point to note is that all of these advertisements and responses happen at the link layer level, without the need to wake the higher-layer host, although proceeding further with the information flow will require host intervention.

7.5 The Bluetooth low-energy state machine

These seven packets define the way in which Bluetooth low energy works. There are other link layer packets, including all of the ones for setting up a link and transferring data, but they are merely turning the wheels once everything is in motion. Concentrating on these seven, we can make sense of the state machine which governs the behaviour of every Bluetooth low energy device. This is shown in Fig. 7.5.

Most Bluetooth low energy devices will spend the majority of their lives asleep in the standby mode. Peripherals and observers will wake up occasionally, typically triggered by their host applications responding to a program or event, or a preset timing agreement with a master when they enter the advertising state or send data using an advertising event. Once they are in a connection, they can go into a deep-sleep mode, but do this within the connected state, waking at the times determined by the master when they first entered the connected state.

Central devices are directed by their hosts to enter either the scanning or initiating state; the latter when they are looking for a device with which to make a connection. The topology for a connection

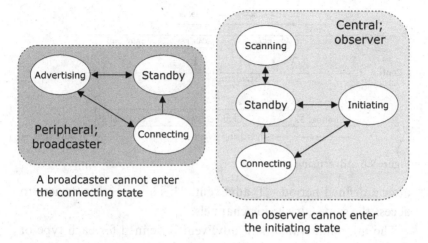

Figure 7.5 Bluetooth low energy functional states

may not always be obvious. Although masters are normally powered devices, there are applications, such as light switches and TV remote controls, where it may make sense to reverse the nomal topology. As a guiding rule, the slave device should be the one where the state information resides, or is applied. In the two examples of light switch and TV remote, those are, respectively, the light bulb (or socket) and the TV set.

Multiple instances of the link layer state machine are allowed, with certain restrictions on the combinations of roles. Hence it is possible for a master to continue scanning whilst in a connection.

7.5.1 Advertising

Advertising events involve sending advertising packets. As the advertiser does not know on which of the three advertising channels a scanning device will be listening, it normally repeats the packet on each of the three channels in turn. In between each packet being sent, it will listen for any requests from a peripheral device that is in the scanning or initiating states. As soon as it has sent its set of advertisements, the advertising device will close the advertising event. It may then repeat the advertising event or start a new event

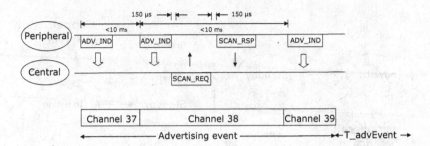

Figure 7.6 Advertising events

after a defined period – T_advEvent. This allows other devices to access the same advertising channels.

The minimum value of T_advEvent is defined for each type of advertising packet. Advertising is a power-intensive operation for a low-power device, so timings of advertisements are important to ensure the minimum power consumption.

A central or observer device may spend most of its life in standby mode. It can be instructed to enter the scanning state by its host controller and, if it is a peripheral, it may also enter the initiating state. In both states it will listen for advertisements. As its receiver is active in both of these states, it will draw power continuously. However, the majority of devices in these states have more robust power supplies. If a peripheral or observer device is battery-powered, care needs to be taken over the duration and timing of its entry into these modes.

Bringing all of this together, Fig. 7.6 illustrates an advertising event using ADV_IND packets.

In this example, the advertiser is a peripheral device sending out advertisements. It starts on the first advertising channel – Channel 37. It receives no responses, so repeats the advertisement on Channel 38. At this point, a central device in scanning mode requests more data using a SCAN_REQ, which is sent 150 µs after it receives the ADV_IND packet. The peripheral responds with the requested data 150 µs later. All these packets are sent on the same channel. The peripheral then repeats the ADV_IND packet on Channel 39 to complete its advertising event. Within an advertising event, each

Table 7.1 *Advertising packets*

Packet type	T_advEvent (ms)	Directed	Allowable responses	
			SCAN_ REQ	CONN_ REQ
ADV_NONCONN_ IND	>100	No	No	No
ADV_DISCOVER_ IND	>100	No	Yes	No
ADV_IND	>20	No	Yes	Yes
ADV_DIRECT_IND	<3.75	Yes	No	Yes

consecutive packet must be sent within 10 ms of the previous one. An advertiser does not stop advertising after a SCAN_REQ. With the exception of the ADV_DIRECT_IND packet, all advertisements are undirected, which means that they are promiscuous – they can be heard by any scanner. The only response that will prematurely terminate an advertising event is a connection request.

Table 7.1 provides a summary of the features of the different advertising packets.

7.5.2 Connecting

To make a connection, a central device must be placed into the initiating mode by its host application. In this state it listens for an ADV_IND or ADV_DIRECT packet, which informs it that an advertiser is available for connection. On receiving this packet from a device it wishes to connect to, it returns a CONNECT_REQ packet. The advertiser immediately stops its advertising event and uses the information in the CONNECT_REQ packet payload to jump to the requested data channel to continue the connection sequence. The two devices will then exchange information and the master (central) device can configure the behaviour of the slave.

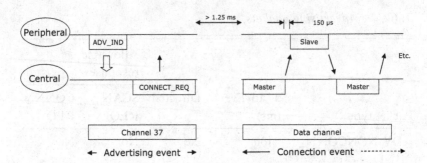

Figure 7.7 Connecting to an advertiser

Having sent the connection request, the master device does not wait for any acknowledgement from the advertiser. Both devices assume that once the connection request has been sent, the connection exists. So after a short delay, the master will immediately start sending packets to the slave.

The advertising and connection process is illustrated in Fig. 7.7.

Once a connection has been established and security information exchanged, a master and slave will typically instruct the slave to enter a low-power mode. However, the connection state is maintained. The initial connection request tells the slave how often the master will initiate connection events (its connection interval) and how many of these a slave is allowed to sleep through before it must wake up to listen for one (the slave latency). If the connection is lost or terminated and the slave wants to re-establish it, it can use the ADV_DIRECT_IND advertising packet to accomplish that.

The ADV_DIRECT_IND packet is the only advertising packet that contains the address of the target device. It is designed to accomplish fast reconnections. Unlike other advertising events, those using this packet can be sent continuously. The trade-off with this fast reconnection mode is that it results in a high-energy consumption in the advertising device. The assumption, when using this, is that the master will be within range and in initiating mode. If it is not, then rather than repeating the process, the slave should resort to the more power efficient ADV_DISCOVER_IND packets.

7.5.3 Discovery

To those used to the terminology of Bluetooth, the discovery process in Bluetooth low energy may be confusing. Discovery is always done by the master device, which is instructed to look for available devices. It does this by entering the scanning or initiating states and waiting for advertisements.

Peripheral devices can be configured to advertise their presence throughout their operational life (general discovery), or for a more limited period (limited discovery). Typically this will be for a short time, either when they are turned on, or a user intervention places them in this state. This difference is denoted by an optional flag within the advertising packet, which allows the scanning device to discard advertisements from devices that are not in the appropriate discovery mode. For example, if it knows that the device it is trying to connect to is set to be in limited discovery mode, it can decide to discard any advetisements it receives from devices that are in a general discovery mode. Limited discovery allows a device to be found faster, and also means that fewer found devices need to be displayed on the user interface.

A master device may passively scan for devices as a background operation. More frequently, it will be directed to look for devices by a user input. In this case, it starts scanning and listening for advertisements. Each advertisement contains the device address and may contain additional data, so that, at the end of the discovery process, the master will have this list for all the devices it has 'found' (Fig. 7.8).

Often a master may want to know more about a device, such as its name, so that a more useful set of information can be provided to a user interface. To obtain the name of a device, the master needs to make a connection to each device in turn and request the additional data. The application can decide whether to do this with some or all of the devices it has found.

As well as basic measurement data, the payload of an advertisement may contain a number of standard pieces of information,

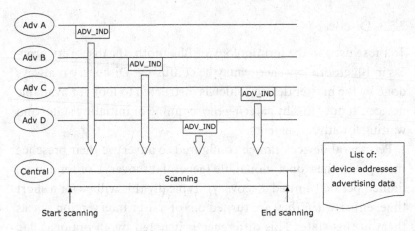

Figure 7.8 Performing a discovery

which will aid the interpretation and display of information to the
end user. These include:

- A local name. If provided, it may remove the need for name
  requests. However, it cannot be greater than 30 octets, and this
  will decrease if other information is included in the advertising
  packet.
- Flags indicating whether the device is in general or limited
  discovery mode. Typically this is one octet.
- Transmit power level. One octet of information. This can be
  used in conjunction with an RSSI measurement, giving the
  master a good estimate of the proximity of each discovered
  device. This information can be provided to the user to help
  them differentiate between devices.
- Manufacturer-specific data.

The total size of the advertising packet payload is 31 octets. Each
of these pieces of data shares this space and each needs to use
one additional octet of the payload to define the type of data that
follows. Hence if the flags and transmit power level are included
(using up two octets each), only 27 octets are available for all other
information.

7.5.4 Bonding

Once a connection has been established, the master may set up a bond, which involves the sharing of keys for encrypted communication. These can be stored for use in future connections. The detail of Bluetooth low energy security is described in Chapter 3.

A server may require authentication and/or encryption for access to some of its characteristics. Where encryption is enforced, the resulting MIC will take up 12 octets of the data payload. Encryption may also be applied to the broadcast data. However, this implies that a previous bond has occurred with the device which will receive these broadcasts. Broadcast profile role devices that have no receiver cannot send encrypted data, unless some out of band method has been used to exchange link keys.

Bluetooth low energy also supports privacy, in the form of private addresses. This feature allows devices that are connected to each other to generate new identities every 15 minutes, which appear as new addresses. These new addresses are composed of three parts:

- A bit to signify whether it is a public or private address,
- A random number,
- A hash of that random number, with some identity information.

Two devices that have previously connected can use their stored connection information to interpret this and resolve the address. Any device that is listening for wireless traffic would only see the 'new' address and be unable to correlate that with previous transmissions from that device. This technique helps to prevent mobile devices being tracked.

7.6 The Bluetooth low-energy protocol stack

The Bluetooth low energy stack is much simpler than that of Bluetooth. It is designed for transmitting the state of devices, rather than streaming data or files.

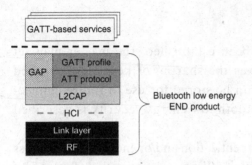

Figure 7.9 Bluetooth low energy stack

Figure 7.9 illustrates the Bluetooth low energy stack. Compared with the Bluetooth BRE/EDR stack, the biggest difference is the lack of multiple alternative transport protocols.

The previous section has covered most of the functionality that resides within the link manager. In practice, developers don't deal with individual advertisements, but access the Bluetooth low energy functionality via the generic access profile (GAP), which is used for discovery and connections, and the generic attribute profile (GATT) to configure devices and transfer data. The GATT profile uses the underlying ATT (attribute protocol) to provide the protocol connection to the lower layers.

As with Bluetooth, an HCI (host–controller interface) provides a standardised API above the link manager, allowing implementers to mix and match controllers from different silicon vendors. The HCI interface is an evolution of the Bluetooth HCI interface, but with a much reduced command set to make it suitable for single-mode chips. The relative simplicity of Bluetooth low energy may mean that the HCI interface is not widely used, as most manufacturers are likely to provide complete system-on-chip designs.

Above the HCI interface, an L2CAP (link layer common adaptation layer) provides a common interface layer for protocol stacks that support both basic rate and Bluetooth low energy. For Bluetooth low energy, there are no connection-oriented channels. Instead of using dynamic channels, L2CAP uses a fixed channel. The default channel MTU (maximum transmission unit) is 23 bytes, although higher values, up to 517 bytes, can be negotiated between clients and servers using the ATT_MTU attribute exchange command. If larger MTUs are requested, they will be broken down by L2CAP and sent as individual 27-byte payloads over the air, which are reconstructed at the far end of the link. So although increasing the MTU allows more information to be sent, it does not make the radio transfer more efficient; hence the default value should be used unless there is a need for longer data strings.

7.6.1 Attributes – exposing state

Bluetooth low energy is designed to transfer state information. It does this using ATT and GATT, which are the cornerstones of the Bluetooth low energy architecture. They signal a major change in philosophy, not just by limiting the protocol options, but by including far more of the data protocol and data format definitions used by higher-level applications. They also define attributes, characteristics and services, which bring interoperability to Bluetooth low energy.

Using GATT and ATT, data are represented as 'attributes'. These may be measurements that are transmitted, or the setting of actuators on a server by a client device. Unlike Bluetooth BR/EDR, there is no other route through the stack. To use Bluetooth low energy, it is important to understand attributes and how they are assembled into characteristics and services.

An attribute is essentially a piece of data. It can signify the contents of a hardware register, device information, a sensor measurement, control data or configuration information. It may be an output from a sensor or an input that controls something.

Every attribute consists of three elements. First, it is uniquely identified by a 16 or 128-bit UUID. This is generally defined within a higher-level profile and identifies what the attribute represents. That may be the physical entity, such as temperature or a type of actuator. Sixteen-bit UUIDs are used to describe commonly used attributes and are listed in the Bluetooth list of assigned numbers, while 128 bit UUIDs are available for the creation of other attributes, which may be proprietary or infrequently used, and hence do not justify their inclusion in an assigned numbers list.

The second defining element of an attribute is its handle. The handle is unique and defines the attribute's location within the attribute server on the peripheral device. Handles are two bytes long; their use allows much shorter messages to be used. They impose a limitation of a maximum of 65 535 attributes within any one device, but this is unlikely to be reached in any practical device (handle 0x0000 is reserved). Multiple instances of an attribute UUID can exist within a server, but each instance must have a distinct handle. Handle values are not allowed to change during the lifetime of a connection. In many devices, handles may be permanently fixed at manufacture.

The final element is the value that represents the contents of the attribute – typically a measurement value, or control information. The maximum length of an attribute payload being written is 20 bytes, and 22 bytes if it is being read. (The discrepancy is because 2 bytes are needed for the handle in a write command.) Longer values (typically strings) of up to 512 bytes can be supported by the use of long attribute values. (The limitation is actually (ATT_MTU^{MAX} – 3 bytes), but most implementations supporting greater MTU sizes accommodate the maximum value of 515 for MTU^{MAX}.) Long attributes are accessed using the read or write blob (binary long object) commands, which access the attribute in consecutive reads.

An implementation may optionally set permissions to define who can access each attribute.

7.6.2 Attribute PDUs

The attribute protocol defines six ways that a client and server can interact with each other using attributes. They are used for all attribute transactions. They are:

- Requests – messages sent to an attribute server, which invoke responses,
- Responses – messages sent to an attribute client in response to a request,

- Commands – messages sent to an attribute server, which do not elicit a response,
- Notifications – messages that can be sent unsolicited to an attribute client,
- Indications – messages that can be sent unsolicited to an attribute client, but which invoke confirmations,
- Confirmations – messages sent to an attribute server in response to an indication.

Attributes only exist on a peripheral device within its attribute server. An attribute client resides on the master (the central device) but does not contain attributes. Although it is the master, it can only send requests and commands to a server (which can also configure its future behaviour), or wait for indications and notifications to be sent to it.

7.6.3 Notifications and indications

Many low-power sensors will operate asynchronously to their clients, reacting to external events, rather than on the basis of a timed response. Typical examples are light switches, pedometers and thermostats. These devices react to an external stimulus and send a message to their clients.

Bluetooth low energy provides three means for sending this information: as a broadcast, a notification or an indication. Broadcasts are promiscuous and use the advertising channels. Indications and notifications use data channels between connected devices.

The form in which the information is sent when using indications or notifications is essentially the same – the difference is in the response. An indication requires a confirmation that the message has been received by the client; a notification does not.

Notifications are designed for situations where battery life is critical and it is acceptable for occasional data to be lost, or where there is another form of feedback. An example of the latter is a TV remote control. These provide 'out-of-band' feedback that the relevant action has occurred.

Indications require a response from the attribute client. Indications are an atomic operation, so no further messages can be sent from the server until the response has been received. If no response is received within 30 seconds then the server will assume that the client is not present and terminates the link.

Both notifications and indications generate baseband-level acknowledgements, so the slave device will be aware that a notification has been received. An indication initiates an acknowledgement from the host stack, increasing the confidence that the master device has both received and processed it.

7.6.4 Characteristics

The attribute is the raw piece of information. Its UUID defines explicitly what it represents and the format of the data. A characteristic builds on this by adding behaviour to the attribute, determining how the information contained within it will be used.

Each characteristic is defined using a characteristic declaration, which itself has a unique identifier (UUID). This declaration includes a characteristic properties field, which provides a mask for the different behaviours. In most cases, devices will be shipped with a default behaviour, which can be modified by the client once it has a connection with the peripheral device. Table 7.2 lists the most common behaviours.

Multiple feature bits can be set which determine the behaviour of the characteristic.

7.6.5 Aggregate characteristics and time stamping

Often, an application may have a number of related measurement attributes. Common examples are longitude and latitude in a GPS device, or diastolic and systolic pressure in a cuff to measure blood pressure. Rather than requiring these to be sent as separate measurements in separate packets, Bluetooth low energy allows these

Table 7.2. *Bluetooth low energy characteristic features*

Feature	Description
Broadcast	Permits the characteristic's value to be broadcast
Read	Permits the characteristic's value to be read by the client device
Write without response	Permits the characteristic's value to be written by the client without generating a response
Write	Permits the characteristic's value to be written by the client, with the peripheral sending a response to the client
Reliable write	Permits a reliable write of the characteristic value, where its value is confirmed before it is written
Notify	Allows the peripheral to send a notification, which does not require a response
Indicate	Allows the peripheral to send an indication, which requires a response from the client

types of related data to be optimised by combining them into an aggregate characteristic.

Aggregated values are defined within a characteristic definition, which specifies which attributes are included within the aggregate characteristic by specifying a list of their handles. When the resulting aggregate characteristic is read, it contains a sequence of the values of each of those characteristics.

Aggregates can also be used where a data value needs to be time-stamped, in which case they will consist of a time characteristic and one or more measurement values.

7.6.6 Services

Characteristics define individual pieces of data. These are collected together within services to provide a set of characteristics that represent a typical function. Simple examples of services are battery status, time setting and thermometer.

Services are put together in a hierarchical manner, so that they can be reused by other services. This allows the concept of a service including another service.

The advantage of this structure is that it allows clients to determine easily whether they can support a Bluetooth low energy server device, without the need to enumerate all of the characteristics.

7.6.7 Configuring attribute servers

When two Bluetooth low energy devices connect to each other, the first task for the client is to discover which features the attribute server supports. It performs this by discovering the services available. Starting at the first handle, clients will enumerate the services and characteristics resident on the server device. Having acquired this information, the application within the client can decide whether any of these need to be configured and which ones it wishes to interact with in the future. After any configuration of attributes on the server, including setting the behaviour for notifications and indications, the peripheral will typically go into a low-power state.

Applications that only support a specific service can minimise the length of this phase by searching for the specific service for the application they support, using the 'request by type' command, which searches for a specific UUID.

7.7 Profiles

Bluetooth low energy profiles differ from those of most other standards. Because there is only one protocol – ATT, which is used for all transactions, and a defined set of commands available within GATT, Bluetooth low energy profiles are limited to defining which services and characteristics are required within the attribute server on a peripheral device and how the client interacts with them. This means they are much simpler: in size (number of pages), implementation, testing procedure and understanding.

Bluetooth low energy profiles, (sometimes also called GATT-based profiles); adhere to an object-oriented structure. Each profile on a server has its own UUID, as do all of its constituent services and characteristics. Once a client application has identified the existence of a profile or service, it knows all of the mandatory characteristics that will be present within the device, and the ways in which they can be read and written. It allows client applications to be written that are free to interact with all of the services, or with any subset of them, giving great flexibility to application developers.

7.7.1 Proximity

Some of the features of Bluetooth low energy make it particularly suitable for proximity applications. These are applications where two devices regularly check the strength of signal between themselves, using their RSSI measurement abilities, along with the known transmit power, which is indicated in the advertisement packets of the server, to estimate how far away from each other they are. Although this is not a quantitative measurement, it can be performed with sufficient reliability to enable a range of useful applications.

This capability of Bluetooth low energy can be used to check when devices are moving out of range of each other. Typical applications involve a tag, which acts as the security input for a phone or a PC. When it is close to the device, it allows it to be used; when it is out of range, the device would be locked. It can also use the range information to sound an alarm, to signify that a phone has been left behind or stolen. The same tags can also be used as secure access or ID devices.

7.7.2 Gateways

One of the most important concepts of the Bluetooth low energy standard is that of the gateway. This extends the normal short-range connection paradigm to define the way in which a device can talk directly to an Internet application.

Gateway functionality requires a central device to implement a generic application. Its task is to provide a secure tunnel to a remote IP address, which hosts an application that works as an attribute client.

To use this feature, the peripheral device must implement the gateway characteristic. This has an assigned UUID and contains the IP address of a web application that supports the device. When the peripheral device is enabled, it advertises that it is looking for a gateway. In most cases, the gateway will be the user's phone. The user needs to accept this request from the device, at which point it will run the gateway application.

The purpose of the generic application is to set up a tunnel to the address held within the gateway characterisic. Once this is established, the remote application residing at this IP address will enumerate the server device as if it were taking part in a local server–client connection. From this point on, whenever the server device has data to send, it will attempt to find the gateway device, and if successful will connect to its IP address and send its data to it. If the gateway is out of range, the server can elect to store its data and then try again later.

The significance of the gateway feature is that it allows designers to develop devices that are intimately connected to a web application. This simplifies data collection, as it no longer requires user input, providing automatic data collection. As the gateway functionality is generic, it means that devices can be built that can expect to connect to the web without the need for users to load drivers or software into the gateway device.

This scheme also simplifies application downloads from an application store to a handheld device. During the initial connection, the client discovers exactly what the device is. It can use this information to search a connected apps store, so that it can automatically display the applications that are compatible with the device. This gives the user the confidence that they will work, and removes the need to look through hundreds of possible applications. Manufacturers can enhance this user experience by including characteristics within their device that direct the user to a specific website to download applications.

7.8 Single-mode chips

As we have seen above, Bluetooth low energy is a completely different and separate specification from Bluetooth. However, the design is such that the two different standards can be combined and implemented within a single chip, with no significant extra silicon real estate. These are called dual-mode chips (Fig. 7.10).

I've already described how the radio section can be shared. The architecture of the Bluetooth dual-mode protocol stack allows both stacks to be combined.

To provide concurrent support of both standards, the dual-mode stack combines both Bluetooth and low energy stacks, using the L2CAP layer to provide a common layer for both. Below L2CAP, the HCI layer interfaces packets with the respective basebands and radios.

Above L2CAP, independent stacks exist, channelling data to the appropriate profile and application. Note that Bluetooth profiles are currently written either for BR/EDR or low energy. They are not transferable.

Dual-mode chips are not combination chips, as is the case with Bluetooth and Wi-Fi combo chips, which are essentially two separate systems on a single die. Instead a dual-mode chip is one where the two radios – Bluetooth and Bluetooth low energy – share the same RF chain within the chip and the two stacks are integrated and implemented within a single processor and protocol stack. Because so much is shared, the cost of these dual-mode chips is no greater than that of existing single-mode chips. (In fact they may be cheaper, as their emergence has coincided with a reduction in process geometry from most suppliers, resulting in a smaller silicon die.)

These dual-mode chips are appearing in the latest generation of mobile phones and laptops. That means that Bluetooth low energy

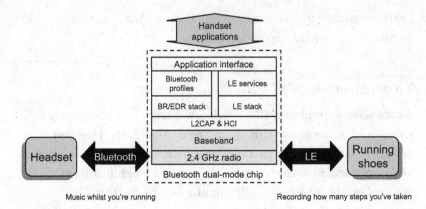

Figure 7.10 Bluetooth low energy dual-mode stack

will be supported by billions of handsets, which will be compatible with both Bluetooth and a new generation of Bluetooth low energy peripherals.

Dual-mode devices can also support the alternate MAC/PHYs described in Chapter 4, allowing them to implement 802.11 over Bluetooth at the same time as Bluetooth low energy.

7.9 Dual-mode chips

Dual-mode chips will be incorporated into mobile phones and PCs, as well as devices like set-top boxes and other devices that are powered and want to act as gateway devices. Most personal-device manufacturers will use low-cost, single-mode chips, which only support the Bluetooth low energy features that have been described in this chapter.

Because of the need to support a diversity of applications, these will incorporate a wide range of features. For the most power-critical applications, there are highly optimised transceivers that will run with separate low-power microprocessors. More commonly, designers will employ general purpose, single chips, which include the transceiver and a host processor running the Bluetooth low energy protocol stack. For the most integrated designs, versions of these chips are available with additional application processors or virtual machines, enabling the complete application to run on a single chip.

7.10 References

[1] The ANT Alliance, www.thisisant.com/.

[2] Original proposals to the IEEE 802.15.4 working group. www.ieee802.org/15/pub/2001/Jul01/. The most interesting documents reflecting the genesis of Bluetooth low energy are: 01230r1P802–15_TG4-Nokia-MAC-Proposal1.ppt and 01231r1P802–15_TG4-Nokia-PHY-Proposal1.ppt.

[3] Bluetooth Special Interest Group, Bluetooth low energy specification. www.bluetooth.org.

8 Application development – configuration

In this and the following chapter we look at how to get the best out of short-range wireless technology. These two chapters focus on the items of a specification a designer can influence within the constraints of each standard and how those constraints may direct the choice of standard. In many cases, these are the same techniques for all of the standards covered in the preceding chapters.

Back in Chapter 2, I talked about the three key differences you need to understand between a cable and a wireless link:

- Working out what your wireless unit is connected to,
- The fact that latency becomes a major factor, as information may not arrive at the far end of the link when you expect it to, and
- Throughput varies in what can appear to be a random manner.

In this chapter I'll concentrate on how these three issues can be addressed and how they might affect your choice of standard, before progressing to ways of ensuring you get the best performance out of that choice.

A standard should be regarded as a well tried basic framework that provides the advantages of interoperability, reduced cost and faster time to market. Anyone who has used short-range wireless will know that within that framework there are many different possible implementations of each standard, each of which can radically affect the way they perform and how they are fitted to the use case. These lead to implementation issues and decisions that will do their best to trip a designer up along the way.

Building on the overview of the different features of wireless in Chapter 2, I show how these map onto the different standards and then how a design can be optimised to best suit your application.

By the end of these two chapters you should have a better idea of how to decide on the most cost-effective implementation and how to get it right first time.

8.1 Topology

Working out what you're connected to takes us back to topology. We have already looked at the different topologies, and seen how the different standards address them. In some cases, the choice of topology for an application will be obvious. If you need a true mesh standard, then you should start by looking at ZigBee PRO. If you need a high data rate connection to the Internet, use 802.11.

Most applications aren't this clean cut. This makes the choice of standard more complicated, particularly if it's a simple case of replacing just one or two cables. In these circumstances, there may be a number of different possibilities, with the optimal choice being a balance of several features. Topology is always a good place to start, as an incorrect choice here will almost always lead to a less than perfect solution. To help decide which standard is most appropriate, we'll revisit the hierarchy of topologies that we first encountered in Chapter 2.

8.1.1 Cable replacement

Cable replacement is what it says on the box – the use of a wireless link to replace a cable. Often that will be an RS-232 cable linking two serial ports, although it may equally be an RS-422, RS-485, USB or audio cable. The important point is that it is intended to be the exact analogue of a physical cable. In many applications it may even be implemented as a wireless adaptor that plugs into the physical connector socket, either internally, or on the outside of a piece of equipment.

It is easy to forget how good a physical cable is, but it's worth labouring the point. Compared with a wireless link, a cable has almost infinite bandwidth, no latency, consumes no significant power, has no security issues and configures itself automatically as soon as it's connected to the appropriate plug or socket. Its only disadvantages are that you can trip over it, it can be expensive to install and its range may be limited to a few tens of metres.

Any of the common wireless standards can be used for cable replacement. It will probably come as a surprise to learn that none of them support it directly within the specification – each needs some degree of assistance to produce a physical equivalent of a cable. Part of this arises from the need to associate the two separate ends of the connection with each other. Whether this is done at the factory and the two ends are sold as a matched pair, or whether a local interface such as a button is added, it requires the addition of a small application to join them together. This is outside the scope of the standards (with the one exception of Wi-Fi's protected set-up scheme). In all cases, additional work is required on top of a profile or protocol stack. The other item that needs to be addressed is the way in which data are mapped to the physical pins of the connector. That's something I'll cover after we've finished with topology.

So what do we do next? Table 8.1 indicates the trade-offs that come from the different standards. These represent typical market performance of readily available chips. Specialist suppliers targeting niche markets may often be able to improve significantly on particular areas, such as power consumption, so some variations will exist, but it's wise to start with some conservative assumptions about average performance.

Cost is always a difficult item to compare, as new generations of chipset can change price points dramatically. I've resorted to giving a relative cost, taking a Bluetooth v2.1+EDR solution as a reference and comparing the other technologies with this. Rather than basing this on chipset cost, I've used a typical solution cost, as this includes all of the peripheral components for the wireless standard, as well as any additional external processing to implement the protocol stack up to the physical connector. This latter point is the main reason for the apparently high cost of Wi-Fi, which needs an external processor to support a full stack for cable replacement. If

Table 8.1 *Characteristics of wireless standards for cable replacement*

	Range (m)	Throughput	Latency (ms)	Current consumption (mA)	Relative cost	Secure?
Bluetooth ACL	100	<2 Mbps	100	35	1	Yes
Bluetooth SCO	50	64 kbps	10	30	1	Yes
Bluetooth low energy	>100	<50 kbps	3	15	0.5	Yes
Wi-Fi/802.11	<100	<20 Mbps	200	200	3	No[a]
ZigBee PRO	>100	<75 kbps	5	25	1.5	Yes

[a] Security for 802.11 in ad hoc mode is currently limited. It is secure in infrastructure mode.

Table 8.2 *Application abilities of wireless standards*

Applications	Voice	Data	Audio	Video	State
Bluetooth ACL	x	Y	Y	x	x
Bluetooth SCO	Y	x	x	x	x
Bluetooth low energy	x	x	x	x	Y
Wi-Fi	(VoIP)	Y	Y	Y	x
ZigBee	x	x	x	x	Y
	State = low bandwidth, low latency data				

you already have lots of spare processing power available, then this can change that equation.

Security should be borne in mind for point-to-point connections. 802.11's ad hoc mode, when used for cable replacement, does not include the enhancements developed by the Wi-Fi Alliance, so additional higher-level security needs to be added if this mode is chosen, which is likely to add cost and development time. ZigBee PRO requires a trust centre for a secure connection. Using it for cable replacement would require one end of the link to be a coordinator with a trust centre, which means that the implementation will be asymmetric. That imposes a much higher current draw at one end of the link. Bluetooth low energy is also asymmetric in terms of its architecture and is likely to require a slightly higher current at the master end of the link. In contrast, a Bluetooth BR/EDR connection would be symmetrical.

The latency and throughput of the different standards may not be relevant for non-time critical data transfer, but it is for other applications, allowing them to be ranked according to use, which is what is shown in Table 8.2. The distinction between voice and audio is that voice normally requires transmission with no delay. That usually equates to a low bandwidth link that is only suitable for voice and not for music. In contrast, audio refers to the transmission of digitally encoded data, in MP3 format or similar. The processing required at either end of the link for audio processing can introduce some significant delays, which are not acceptable for applications like voice telephony. As always, there are ways around this, but once again, they start to introduce additional costs.

The last column of the table – state – refers to the transmission (or retrieval) of the state of information from the other end of a link. This may be the value of a sensor or programming an actuator. Although these are data, the requirement is frequently low latency, for instance where the data may be a switch closure or alarm signal. It is best served by a network that responds very rapidly to a request to transmit data, rather than an IP-based system, which may introduce delays. Standards designed for transmitting state information are normally low-duty cycle systems, where the sensor end of a link is asleep for most of its life. The best way to think about state is as a register or port at the end of a wireless link. By introducing the concept of state means that the definition of the data column in Table 8.2 can be more limiting. I have taken data to be those applications where significant amounts of data are likely to be transferred, often in the form of files, rather than a piece of state information. Although both ZigBee and Bluetooth low energy can transmit moderate amounts of data, neither are particularly efficient at it and rapidly lose their low-power advantage. Both are designed primarily for the transfer of state.

Cable replacement may not seem an obvious candidate for a wireless standard, as replacing a single cable doesn't benefit from interoperability, as a single manufacturer would normally supply both ends of the link. Despite this, wireless standards are increasingly used for cable replacement, because of the other benefits that they bring: robustness, small size, alternative sourcing and ease of use. As a result, chip vendors and module manufacturers provide a range of options for implementing end-to-end cable replacement solutions. Where a small number of connections are to be shipped, it is usually advantageous to choose one of these off-the-shelf solutions.

8.1.2 Reconnection

Although it's not a problem limited to cable replacement, implementers should think carefully about how to handle dropped connections. One of the other benefits a cable possesses is that it has few failure modes, most of which are visible. Most failures of a cable connection are as a result of it being unplugged, which has an obvious remedy –plug it back in.

Wireless is far more mysterious in that there is generally no indication as to why the connection has disappeared, nor any immediately obvious way of reinstating it. Particularly where wireless is used for embedded devices, which may not have a user interface, designers should consider appropriate recovery techniques to restore a lost connection.

When starting a wireless design, always assume that there will be occasions when the wireless link fails, even though both ends of the link may still be working. Even in the best designed systems this can occur because of interference, movement of one or both ends, taking them out of range or an obstruction of the signal. As well as a fundamental loss of link, there are the possibilities of failures associated with invalid packets that are not correctly rejected by the stack. Such issues shouldn't occur in stable stack implementations, but in the real world they do. For that reason some form of watchdog and recovery process should always be employed within each wireless node.

Failure of the wireless link needs a reconnection strategy. In most cases, cable replacement scenarios assume that, like a cable, the link will be present, so it is important to signal to the host device when a connection is not available and take measures to reconnect it. All of the wireless standards provide a means to check the existence of a link. When a link loss is determined, steps should be taken to reconnect, generally automatically. Depending on the standard, attention should be given to data that may have been transmitted and lost, or queued for transmission. Where a system is employing higher-level protocols, such as TCP/IP, this may already be handled. In simpler connections, such as RS-232, a proprietary protocol used between the devices may cope with these interruptions. However, a lot of these protocols were designed for cables, where the possibility of a link failure was not considered. In these cases it may be necessary to add a more resilient protocol within the products themselves, or consider buffering within the wireless adapters.

A more intelligent approach is to take preventive steps. Most standards provide an indication of the quality of the link, generally by providing access to RSSI information. That allows a local application to be written within the wireless adaptor that looks for changes in the link quality and takes pre-emptive action to warn the host. Where appropriate, it can do this using standard flow-control signals. In its simplest form, this could be an audio or visual alert to a user of a handheld device that it is going out of range. This fits better with some chip implementations than others, as it requires two separate processes to be running – one to check the link and one to cope with the general data flow. That is an implementation issue, not a standards one.

Reconnection strategies add complexity to the wireless adapter design, but provide a much smoother operation. Where wireless is embedded into devices that the user expects to work, it is important to invest time and attention in developing these. Most standards provide tools to help do this, but the specific recovery strategy is normally application-dependent. If automatic reconnection and link-quality detection are included, it provides the appearance of a device that works as well as if it were cabled. Without them it is very easy to end up with a product constantly requiring the user to reset it to make it work, which decreases confidence in the wireless solution.

8.1.3 Multipoint

When additional connections are added, life becomes more complex. Multipoint generally refers to multiple independent connections from a master device to a number of slave or sensor devices.

It is distinct from broadcast, where a transmission from a master is received and potentially acted upon by several devices, which I'll come to in the next section when I talk about time synchronisation.

When embarking on a multipoint design, there are a number of important points to consider.

8.1.3.1 Shared bandwidth

As there is one master device, the bandwidth of the wireless connections for each slave comes out of the total that is available for the master. So if two slaves are connected concurrently, they should not be designed to utilise more than half of the bandwidth.

In some applications, although several devices are connected, most of the time they will be idle. In this case, an individual device may be able to access a greater percentage of the bandwidth. Consideration needs to be given to how quickly other devices may connect when they need to. Here, a trade-off may be made between time of connection (if it is a connection oriented system) and maximum bandwidth while connected.

8.1.3.2 Multipoint connection sequences and latency

Depending on the wireless standard and the way it is implemented, the slave devices may connect in different ways. If connection latency is important, then the master may poll all of the devices to check whether they need to send data or to see if a transaction is pending. Alternatively, devices can signal to the master that they wish to commence communication. With multipoint, there is generally an assumption that one of these techniques is being used, so that devices do not need to go through an association process before starting data transmission. In other words, all of the association and bonding is assumed to have taken place when the devices were first brought together.

That doesn't prohibit a mix of connected and disconnected nodes in a multipoint topology. This may well be desirable where a few devices need low latency, whilst other devices may just wish to signal an occasional event. In general, a connected device is able to send data in a well defined period, giving it a lower latency, whereas a device that disconnects needs longer to reconnect, as that process involves the exchange of a number of messages with the master device to re-establish security. On the other hand, being permanently connected, even when in a low power state, will result in a higher operating current for that node.

This highlights the fact that multipoint topologies quickly add complexity, both in terms of managing the links and the effect on power consumption of individual nodes. The choice of wireless standard and the way in which devices are accessed can have a significant bearing on the power consumption in multipoint scenarios. In general, a slave device will be designed to reside in a low-power state other than when it is requesting attention or it is involved in data transfer. However, that places a higher burden on the master device, which may need to stay in an active mode to check for messages from the slaves.

As the number of devices is increased, the response time of a master device is likely to degrade. Other, unexpected effects can come into play. Standards that employ a listen-before-transmit mode may suffer from interfering transmissions as the number of nodes rises. Connection-oriented standards like Bluetooth don't suffer from this, but pay the penalty of having a limit of seven concurrent connections.

Where data throughput is very low, the approach used by ZigBee and Bluetooth low energy can give very low latencies, but with a limited data throughput. That emphasises once again the difference between file transfer and streaming at one end of the connection scale and event signalling at the other. A single cable does both well. Wireless standards are far less universal and need to be chosen for the specific application.

8.1.3.3 Memory requirements
A practical consideration for multipoint topologies is the amount of memory and processing power the master needs to support more slaves. This can grow rapidly, even with quite low numbers of

devices. For Bluetooth, particularly where the host stack is included within a single chip, performance will degrade noticeably as soon as a third slave is added. Similarly embedded Wi-Fi implementations, using ad hoc connections, can exhibit significant degradation in throughput after three or four connections.

8.1.3.4 Multiplexing

One obvious corollary of a multipoint topology is that the master must cope with data streams from a number of different slaves. These require a multiplexing or addressing protocol within the higher layers of the master device's protocol stack to ensure that both incoming and outgoing packets are delivered to the appropriate service, and that flow control and quality of service requirements are met. Although standards that support multipoint configurations contain the tools for these multiplexing layers, they normally need to be implemented specifically for the application. This can result in a standards-based solution that is dependent on a specific chipset implementation within the master device, along with a proprietary API. Slave implementations should remain fully compliant, as they only have knowledge of a single connection stream.

Most of these considerations disappear for low-duty cycle applications, which are more likely to use ZigBee or Bluetooth low energy. Although the same fundamental issues apply, in these applications the duty cycles are so low that these particular concerns rarely impinge. However, to maintain good latency connection, it is likely that the master device will need to be powered and constantly ready to accept an incoming packet.

8.1.4 Infrastructure (network connectivity)

Wi-Fi or 802.11 is usually the first choice for devices needing to connect to a backbone network. It has the obvious advantage of being part of the same 802.3 standard family as many deployed networks. For client devices with operating systems that already contain a TCP/IP stack and that have an adequate power supply, there's little reason to use anything else. For smaller embedded devices, power consumption may cause a problem.

To address this market, a number of silicon companies have developed low-power chips based on the 802.11 standards, but which are highly optimised to provide very-low-power consumption and correspondingly long battery life. Amongst such companies are G2 Microsystems,[1] GainSpan,[2] Redpine Signals,[3] ZeroG wireless [4] and Ozmo Devices.[5] These chips usually operate at the lowest data rates of 802.11b, have aggressive power-saving modes and integrated application processors with lightweight TCP/IP stacks. They can support sustained throughputs of around 1 Mbps, but are targeted at applications that wake up, send small amounts of data and return to sleep. Their main area of use at present is for asset tracking, where they are deployed as active RFID tags. As Wi-Fi access-point infrastructure is readily available and often already deployed, these remove the need for custom access points or hubs, which are normally required with other wireless standards. Although not as low power as Bluetooth low energy or ZigBee, they provide the advantage that they can be integrated into existing Wi-Fi systems without the need for new infrastructure. For low-volume applications, this may make 802.11 a compelling choice.

One consideration in using these chips is the level of security. The more recent security protocols of WPA and above, particularly in enterprise mode, require significant resources that may not be available without adding an additional application processor, which may increase the power consumption to an unacceptable level. As the Wi-Fi Alliance is constantly upgrading its security schemes and requirements, this may limit forward compatibility with future generations of access points, particularly if they are being deployed as part of a corporate wireless infrastructure. Other wireless standards normally require their own infrastructure, which divorces them from the corporate cycle of updating Wi-Fi security.

At the moment, these low-power chips are 802.11 solutions, not Wi-Fi. They may only operate at a limited number of 802.11 coding schemes, resulting in problems with some access points. This limitation in coding schemes, along with limited support for the latest security modes, means that the devices built using them may not be certifiable to Wi-Fi requirements. In most applications that is not a problem, but it may be a feature that is important to some customers.

Any of the other standards can be used for network access, and are, but each needs its own access points and routers to make the connection. Unlike the case of Wi-Fi, these are neither readily available nor is there widespread installed infrastructure. That may change in the case of Bluetooth low energy, where mobile phones will act as gateways to connect devices to a web service.

8.1.5 Cluster tree

Before we get to true mesh, it is worth pointing out the existence of cluster-tree networks. In terms of this book, they are possibly irrelevant, as the standards themselves, with the exception of ZigBee, don't include cluster tree network support. However, a number of vendors have built cluster-tree stacks that work on top of standards. These are used for large-scale sensor networks and back-haul networks of access points.

These exist because the application requires a specific feature of the underlying standard. In the case of Wi-Fi, it is to provide a backbone network to connect municipal access points, removing the need to provide backbone cable connections in inconvenient locations. Examples of companies involved in this area include MeshDynamics,[6] Strix,[7] Skypilot [8] and Tropos.[9] Although these are proprietary extensions of wireless standards they address particularly demanding applications.

8.1.6 Mesh

If your application needs mesh then you should look at ZigBee PRO. Although most 802.15.4 silicon vendors can offer their own proprietary mesh network, ZigBee remains the only standard. So if interoperability is important, it is the only choice. As pointed out before, make sure you really do need mesh. Despite the simplicity brought by the tools provided within ZigBee it is still a complex technology for designers and users to understand.

8.2 Data protocols

Wireless standards don't extend very far into the real world of applications. They do what it says on the tin – provide a wireless link between devices. When they are used in products, designers need to find ways to get the standard's interface to talk to their own specific application.

This tends to be a very different process if a completely new design is under way or if wireless is being added as an upgrade to an existing product. With a new design, the wireless stack becomes an integral part of the design. In contrast, when wireless capability is being added to an existing product, the wireless stack needs to be interfaced to a defined legacy protocol.

8.2.1 Profile or proprietary

As we've seen whilst looking at their architectures, different wireless standards stop at various points up their protocol stacks. Some include profiles that effectively define complete applications, others provide standardised transport layers and emulate physical ports, whilst some rely on drivers to interface with other industry protocols.

There is not a 'best' type of profile. The reason that the standards have so many different profile options is that in attempting to achieve interoperability, they need to take a number of different approaches. These are governed both by the type of application and the diversity of anticipated implementations. For example, a smart energy meter and a headset both need a high level of interoperability within their respective ecosystems, but only a limited amount of product differentiation. Medical equipment requires a high level of interoperability, but here the wireless standards tend to limit themselves to the transport itself, allowing more flexibility in product design. And at the other extreme, general-purpose profiles or transports, such as IP and serial port emulation, provide scope for enormous variations of application, but end up supporting proprietary applications that have no interoperability.

When looking at your application, you need to start by asking about the level of interoperability you need. If your product needs to talk to products from other vendors, then unless you are implementing a pure IP link, you are likely to need an application profile. The more interoperable it needs to be, the tighter the profile requirements are likely to be. So start off by assessing which products you want to communicate with and checking which profiles and standards they use. If you are entering a field where there are few wireless products and no generally accepted standard already established in the market, look to see if there are any industry-specific initiatives that are defining them. Failing that, if there are other manufacturers in your area with whose products you want to be compatible, talk to them about whether they are following a particular standard group. Most industries are discovering that to grow the market and surrounding ecosystem, they need to embrace interoperability. As a result it is no longer considered beneficial to devise your own interfaces if you want to be part of that larger ecosystem.

If you don't need to talk to any other product (and that includes phones, PDAs, access points, printers and PCs – all of which support existing wireless profiles), then you may not need to employ a profile at the top of your protocol stack. Most standards provide interfaces or simple transport profiles that let you write your own proprietary profile or application on top. When taking this route, be aware that you are throwing away the benefit of interoperability and just using the underlying characteristics and robustness of the wireless link.

8.2.2 Interfacing with external protocols

Wireless standards rarely know much about other protocols, particularly the specific wired one that your existing product already has. That means it is necessary to spend time in understanding how to connect them to your own product. Even if there is a suitable application profile, there may still be the question of how to get the data being transmitted into the device.

For most applications that don't have an application profile, designers need to resort to the physical interface provided by the chip or module being used. A few transports defined in some standards relate to ports, such as RS-232 and USB in Bluetooth, but in most cases interfacing the wireless link is up to the chip designer. Even in the case of Bluetooth, these interfaces are optional and may not be implemented on the chip you want to use.

If you are adding wireless to an existing product, then your interface and protocol are probably already defined. If you own both ends of the link then you can retain your existing data formats. If you're planning to interoperate with other devices then you will need to adjust these to meet whatever standard you aim to support.

Where products continue to support their legacy wired connections, care should be taken to ensure that protocol changes are not reflected in the legacy connection and also that the equipment behaviour is well defined for those occasions when users make a wired and wireless connection at the same time.

Many module vendors try to take this issue away by integrating support for standard interfaces within their modules and providing documented APIs for them. As they have done this work for you, they often provide the fastest route to market, although it may not be the cheapest bill of materials. Wireless modules are available with RS-232, RS-422, RS-485, USB, SPI, I2C and USB interfaces, as well as more specialised ones for particular industry areas.

8.2.3 Voice, audio and codecs

Moving beyond data protocols, a requirement to support voice or audio immediately starts to limit the wireless options. Remembering that latency is one of the curses of wireless, the first question to address is whether the requirement is real-time or not.

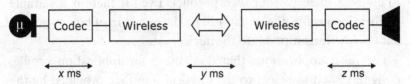

Figure 8.1 Latency in a voice transmission

Real-time means that you want to consume the data stream at the far end of the link at the same time as you are sending it, as you would with a physical cable. This means that there is little or no time for resending lost or corrupted data. As data are continuously being produced at the originating end, there are no gaps to pause and try again. There may be a chance for a few retries, but real-time demands a philosophy that the data has a shelf life of a few tens of milliseconds. If the data cannot be replaced within that window, then they must be discarded.

Lost or discarded streaming data are normally very obvious to the recipient, either as random pixellation on a video display, or noise, pops and distortion on voice or music. For that reason, streaming wireless needs low latency and good quality of service. To provide a good margin of error it also benefits from a basic data transfer rate several times greater than the application require-ment, to give it plenty of overhead.

A follow-on requirement for real-time streaming is that both ends employ real-time codecs. Consider the example of a voice call (Fig. 8.1). As well as the inherent latency of the wireless link, which may itself be several tens of milliseconds, the codecs that encode and decode the voice signal can add similar delays. In the case of video or audio codecs, these may combine to several hundred milliseconds.

This is why Bluetooth uses a relatively simple continuously variable slope delta modulation (CVSD) codec for headsets. The resulting quality is limited, which is why it can only be used for voice and not for music, but the coding delays are minimal.

Moreover, guaranteed times slots using SCO or eSCO are reserved for the link.

The higher bandwidths required for stereo music preclude the use of SCO channels on Bluetooth. Instead, the A2DP profile uses the higher-capacity data channels. These include a range of codecs, which the devices can negotiate at the start of each connection, with a default of a license-free SBC codec. Where music is streamed by itself, latency is not usually a problem. When it accompanies video, which is displayed via a different route, then lipsynch issues need to be considered.

The Wi-Fi Alliance has a slightly different problem, as it has more than sufficient bandwidth, but less guarantee of latency. It has addressed this with its wireless multimedia profile (WMM), which requires the quality-of-service options specified in IEEE 802.11e. It is left to the end application to negotiate the codecs that are used.

The greater amount of work needed to encode music or video to CD- or DVD-like quality requires both more processing power and more memory. That means that power consumption is higher, hence larger batteries are likely to be required for stereo headsets than for voice headsets. Fortunately, the form factor of stereo headsets means that this is not normally a problem.

As codecs become more complex, the processing time increases to the point where real-time operation is no longer supported. In some applications, like two-way voice, this is unacceptable. For music, it depends on the application. Where the requirement is to stream music from a source to a speaker, a delay may not be an issue, as the listener has no external time reference to inform them about the delay. (It may be useful to use a parallel low-latency channel to send control signals, so that there is no apparent delay in response time.) However, if there is a reference, such as when the user is listening to sound on a wireless headset, but watching video on a PC, the delay can become particularly annoying. This problem is known as lip-synch delay. There are two approaches – either to choose a low-latency codec, or to add intelligence to the link so that the video playback can be delayed to cope with the actual output of audio over the link. Although both are being investigated by standards groups, neither is covered in current specifications. The simpler approach is the use of faster coding schemes, and a number of vendors offer these codecs for license. However, if the same higher-quality codec is not available at both ends of the link, it will revert to one of the other codec options, which may cause the lip-synch issue to reappear.

Real-time video offers even more of a challenge. Whereas voice and music codecs may be implemented within wireless chipsets, any attempt to support real-time video is likely to require external codecs, preferably implemented in chips with specific hardware acceleration. Fortunately it is not a common requirement.

8.2.4 Latency and time synchronisation

Latency is not just a problem for streaming – it is very relevant for many applications where actions need to be performed over the link. In the industrial arena, applications that have strict latency requirements include actuator control and reporting of data.

Although multipoint may appear to provide multiple simultaneous connections, in reality these are time multiplexed, with connections being made sequentially to different slave devices. That rarely causes a problem, but if there is a requirement for actions occurring on two or more slaves to be synchronised accurately, then additional protocols may need to be added. None of the standards is specifically designed for this latter scenario, although some do provide synchronisation facilities. Bluetooth's MCAP protocol, which was developed to support synchronisation of multiple medical sensors, is the best option within any of the standards, as it allows time synchronisation between devices down to 10 μs.

Time synchronisation can be very confusing. In wireless, it is normally applied to the technique of synchronising the internal clocks of multiple devices. These clocks can then time-stamp measurement data, so that they can be collated and aligned within a master device or back-end application.

Synchronising actions within wireless nodes is more problematic. If the actions are to take place at some defined point in the future, then command messages can be sent in advance and the synchronised clocks used to ensure that they happen concurrently. If the actions need to be initiated and performed in real time, then the only option is to use broadcast, and hope that none of the slave devices will miss the message.

In almost every case, accurate synchronisation requires application-level programming, which is outside the scope of the wireless standards.

8.3 Set-up and commissioning

8.3.1 Pairing, bonding, association

The addition of wireless means that you will either need to find a way of pairing to the correct device, or else pre-pair devices and deliver them as a matched pair. As a result, it is almost inevitable that the act of adding wireless to a product means that a user interface will need to be rewritten, to provide options for the product to connect to another wireless unit or access point. However simple the connection, that modification should be used to schedule all of the wireless-related firmware changes, including any protocol conversion, rather than vice versa.

In an ideal world, devices would have enough intelligence to know what to connect to and when to do it. Unfortunately they don't, so designers need to add the ability for them to display a list of potential devices to which they could connect and allow the user to decide what to do. They may also want to provide the option for the user to select the security level for the current and subsequent connections. This interface is generally beyond the scope of the standard and relies on the ingenuity of the product designer. It does, however, use tools provided by the standards. From a user perspective, it affects the usability of a product, perception of the brand, and in the case of commercial installations, the cost of ownership of the product. Despite the importance of all of these, it is frequently still done appallingly poorly.

As we've seen before, it's not the standard that dictates how well this can be achieved, as all that the standard does is to provide the toolkit for designing connection schemes. Often the problem is that the user interface is designed by engineers who understand how the wireless connection and infrastructure works and forget that the eventual user does not.

To design a satisfactory interface, the designer needs to consider how the product is likely to be used, where those connections will need to be made and by whom, how often the process will be done and how to recover when it goes wrong.

8.3.2 Promiscuity

Wireless standards refer to connections made in an opportunistic or unstructured manner as promiscuous. These are often non-secure connections, which may only last for a single session. Typical examples may be a laptop connecting to a Wi-Fi access point in a hotel, or a mobile phone making a Bluetooth connection to a friend's phone to send a photo. They're a feature of interoperable standards, and are a key component of the expanding ecosystem of connectable devices that make wireless standards such a popular choice.

Users like this ease of use. At the other end of the spectrum, corporate IT managers do not. Your intended customer base will probably influence how you address connectivity.

Promiscuous connections, with the exception of devices that broadcast information automatically, demand a user interface so that the user can decide whether or not to allow the connection. At the radio level, it is assumed that the radio is allowed to scan for devices it can connect to, present the information to the user and then proceed to connect on the basis of a user command.

This underlying scanning behaviour is a key aspect of most connection regimes. Designers may be able to modify this, based on information that the remote device returns, outlining its capabilities. For example, filters can be added that will only report devices having specific profiles, or even ones that have a known range of device addresses. Adding this level of intelligence can enhance the users' experience, as it does not present them with long lists of devices to which a valid connection cannot be made.

8.3.3 The initial connection

An important consideration is when a product should be allowed to attempt to make a pairing. This is not always obvious and varies between devices. For consumer devices, it may be a process that is initiated the first time a device is turned on, or when a battery is inserted. It may also be initiated by a specific 'pairing' switch, or an action on the user interface. Unless a specific filter ensures that only appropriate devices will respond, it implies a second user interaction that will select the correct device after a list of devices within range has been presented.

A first rule of thumb in designing the user interface is to apply intelligent filtering and only display those devices to which it makes sense to connect. For a ZigBee network, it probably doesn't make sense to display thermostats when commissioning light switches. Nor does it make sense for a Bluetooth printer to detect and display keyboards. The concept of wireless connectivity is still alien to most users: displaying irrelevant possibilities will only confuse. The endpoint device ID in ZigBee and the device icon in Bluetooth, if used sensibly, can help designers to make sense of the multitude of connection possibilities in their user interfaces.

It is possible to configure most wireless products so that an action is required at both ends of the link to initiate a connection. This is generally a sensible approach, as it protects against unauthorised devices trying to make connections. When implementing schemes like this, bear in mind that they are generally proprietary applications that you have built on top of a standard, so may appear inconsistent to a user. The manufacturers of devices you may want to interoperate with will probably have made different choices. Don't do something different from the rest of the industry without good reason, as differing interfaces will confuse the user. Don't underestimate the power of what they have already learnt.

One of the simplest techniques is to have a 'connect' button on both devices. When both buttons are pushed at the same time, the devices connect. Even this is moderately counterintuitive to most users, and rarely works between devices from different vendors. The more complex connection schemes that are frequently employed often seem designed to create a support call. Whatever scheme you choose, test it on potential customers at an early stage of the design process. A good example of an attempt to standardise the simple button approach is the wireless protected set-up specification from the Wi-Fi Alliance.

The more that can be done to remove decisions from the user during the connection process, the better. However, in achieving this, there is a potential trade-off in usability and security, so these should be considered as part of the connection interface design.

8.3.4 Out-of-band techniques

The initial connection process does not need to involve the wireless link. Out-of-band (OOB) connection refers to the use of another technology, which is used to transfer connection data and usually link keys between two wireless devices. It is termed out-of-band because the technique involves something other than the wireless standard. As well as making the connection easier for the user, it can provide an additional level of security, by removing the opportunity for man-in-the middle attacks.

Popular OOB approaches include the use of near-field communication (NFC), which is triggered when two devices are touched against each other. The NFC approach can also be extended to a keyfob-type tag which is touched to a number of devices, providing pairing keys to connect the set.

In devices with connectors, OOB pairing can be achieved by writing an application that transfers the connection data when the two devices are physically connected with a cable.

Another popular approach is to use a barcode scanner. A device containing a scanner scans a label on the other device, which contains information that is used to connect them.

The advantage of many of these OOB techniques is that they are so easy to use that the connection can be remade every time. Whilst that may seem counterintuitive, if the products are not owned and used by a single user, this allows them to be picked up and connected from a central store by a number of employees. This can save considerable time in stores or factories where devices like barcode readers, handheld printers or credit-card readers are used as a communal resource, rather than being used exclusively.

A number of companies have successfully employed this approach, where all of the wireless devices are returned to their chargers at the end of the day. This action deletes all of the pairing information within them. The next morning, the employees pick up any two devices at random and pair them afresh by touching them or scanning each other. The employees do not need to know that they're performing a wireless pairing – they just know that if they do this, it works. The technique has a secondary advantage that because the connections are not permanent, damaged or faulty devices can easily be swapped out without the need to reconfigure other devices that connect to them.

8.3.5 Disconnecting

Many devices do not want to keep the relationship with each other for ever. It may be because they fail, or because they are replaced at different intervals. In either case, it is necessary to make new connections. When that happens, it is good practice to remove the old connection information and any stored security keys.

Retaining connection information has an impact on memory requirements, often at chip level within the MAC or link manager.

That means devices that make many connections may find that they run out of available memory and fail to make new connections. This is another reason for limiting the amount of historic connection data that are stored. A simple approach is to store data for the most recent connections and purge older connection data. The approach of connecting each time, or each day as described above, is an alternative and very effective way to approach this problem.

8.3.6 Limiting broadcasts

Some devices are designed not to connect but to broadcast data for other devices to hear. In some advertising applications, they also attempt to make a connection to a device and push data to it. This normally requires the user to opt into the service and set an appropriate security level.

For this type of application, it is advisable for the broadcast device to keep a list of device addresses it has managed to connect to, so that it does not send multiple messages to the same device. Where multiple broadcasters are deployed in one location, consideration should be given to networking them and sharing this database of previously paired devices to limit the degree of customer disturbance.

8.4 Feature creep

Wireless seems to induce feature creep more than any other technology. Unfortunately, it is the one most likely to be broken by constantly changing requirements. So the rule should be: decide on your design specification and set it in stone. The worst possible area for feature creep is the RF section. If it works, do not change it.

Always remember that your aim is to get a product to market, not to redesign a specification. Work with the most up-to-date stable version of the release unless there are features in newer releases that you absolutely must have.

8.5 Security

Never ignore security. As I have emphasised several times, wireless does not have the inherent security of a cable. Instead, authentication and encryption need to be employed. The media love to publish stories about personal data being stolen, and poor security in a wireless product can result in adverse reportage that can kill a product's sales, or even those of a complete technology. An equally important consideration is the resulting product liability in the event of a link being compromised and customer data or actions stolen or corrupted. On both counts, designers should ensure that the security of a wireless link is fit for its purpose.

All of the standards place great emphasis on their security and all have robust implementations. Always use the most recent stable version of the standard and implement all of the applicable security features. Critical applications should also consider adding further end-to-end authentication and security processes at the application level.

8.6 Upgrading

Wireless standards change. So do your applications, unless you're in the privileged position of writing bug-free code and having customers who are always satisfied. As a result, an important consideration in the process of specifying your product requirements is whether you believe it will be necessary to upgrade your products after you have shipped them. Be aware that upgrades, particularly over the air, have the potential to stop products working. Unless you are absolutely sure you need this capability, it is better not to include it. If upgrade capability is included, it is likely to increase cost; to perform reliable upgrades there is an impact on memory and processing requirements. It may also lead to degraded performance, depending on how much of the system bandwidth is consumed by the upgrade process and how often it is performed.

If you are in doubt about the need, the best answer is generally not to implement upgradability. Again, select the most recent, stable version of the standard and stick to it, resisting any attempts to persuade you that there is something newer and better. Where wireless is involved, specification creep is always best avoided.

If you decide to support upgrades, that decision will help to determine your choice of chipset provider and architecture. Even in the worst case, the need to upgrade should be very limited, with no more than a few upgrades during the product's life. The exception may be where an application residing on the wireless device needs regular upgrades or updates, such as when a database needs regular updating.

In theory, the choice of standard is independent of whether or not you need to upgrade, but it can influence the resources you will need for a robust upgrade process and hence the product cost.

Different standards have different-sized stacks. There is generally a more or less linear relationship between the size of the protocol stack and the data throughput of the radio link. So Zigbee has a relatively slow data throughput, but a moderately compact stack, whereas Wi-Fi has a much higher throughput, but a correspondingly larger stack.

The application sitting on top of the stack, or the size of data files used by it, may be much greater. For a successful upgrade strategy, these may need to be sent to all devices. For networks containing many different devices, the upgrade strategy needs to ensure that the correct files are deployed to the correct devices in the correct order. Take care to understand what load this imposes on the network. I have come across upgrade strategies that involve sending so much data that the network never completes the upgrade. That's a particular concern where the topology of the network means that the upgrade needs to be sent over multiple elements of the network. With good design, neither should cause a major problem for any wireless network, but the data load should be considered at an early point in the design, as it impinges on the choice of wireless standard and network topology.

That brings us to the choice of chip. If you want to upgrade firmware, don't go for a ROM-based design. There is a compromise available, where some manufacturers may implement the lower layers of a stack in ROM, but the higher layers in RAM. This is generally safe, as changes to the lower levels often require a hardware change as well.

The most important consideration is to make sure that your design can recover from a communication or power failure in the middle of an upgrade. If it can't, the customer has a dead unit, which you will need to replace.

For a safe over-the-air upgrade, there are two approaches. One requires a chip specifically designed for upgrading; the second requires additional intelligence to be added to the product to do this, usually in the firm of an external microprocessor.

Chip-based approaches typically contain either enough memory to hold two images of the firmware, or else flash memory for one and a ROM-based version that can be switched back in if the upgrade image fails. When an upgrade is performed, the new image is stored in flash and then loaded into RAM the next time the chip is powered up or reset. If the resulting power-up fails to work, this is sensed, the original image is loaded and the chip restarted. An error condition is then sent to the host, to inform it that the upgrade has failed and should be attempted again. Because this involves more memory within the device, to cope with storing the incoming image as it is written into flash, supporting it is not a popular option with chip vendors as it increases chip cost. They prefer to push this cost to an external flash memory.

That means that most designers will end up implementing the upgrade process externally to the main wireless chipset. This normally involves an external application processor and using its associated flash memory for the upgrade process. Here, the main concern is to ensure that there is a fallback position in case the upgrade process fails. The steps to achieve this are shown in Fig. 8.2.

- Start with a design that includes two firmware images: a known working image, which is normally the image that is shipped with the product, and memory space for an upgrade image.

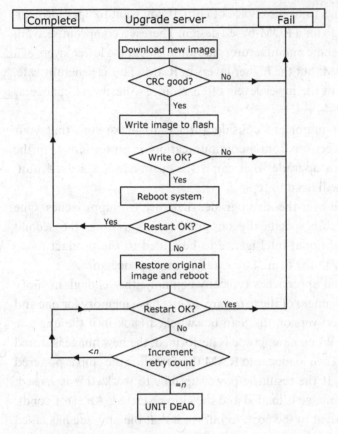

Figure 8.2 Upgrade flowchart

- To perform an upgrade, download a new file to the external processor and store it in the memory space for the new image.
- Once it has downloaded, check the integrity of the image. If it does not meet the check condition, discard it and initiate another download. If the download fails at any point, reject it and restart.
- Once the image is confirmed, run an upgrade routine from the application processor to write this to the flash memory for the wireless chip and force a reboot of the wireless chipset.
- Firmware images should always be written so that on successful power-up they send a confirmation signal to the application processor that the wireless chipset is ready. If this is not received within a reasonable time, the application processor should

upload the original image to the chipset and force another reset. This should restart correctly, at which point it can report an error to the upgrade host and request a new image.
- If successful, the host may decide to copy this as the new default image. (Be extremely careful about doing this – it may introduce as yet unknown faults in future upgrades.)

Although complex, the point of the procedure is to ensure that the device can never be left with a non-working firmware version and hence be incapable of connecting via the wireless connection. It requires extra resources, which adds to the cost, but it is safe. Taking a simpler approach runs the risk of a dead unit.

In the case of systems that have partitioned the stack and application across several processors, a more complex version of this procedure may be required to update each portion. In these cases, care must be taken to ensure that the individual parts cannot reboot with incompatible versions of software or firmware in the wireless chipset and application processors. This requires particular care to be taken if there is a need to update the operating system on the application processor.

At the end of an upgrade process, a wireless network should continue working in the same manner as it did before the upgrade. To do that, make sure that all of the operational data in each node, such as the routing tables, pairing data, passwords and link keys are securely stored and survive the upgrade process. Over the years, I have come across a number of upgrade implementations that have overwritten them or returned devices to a factory default setting as part of the process, leaving an upgraded, but disconnected network.

8.6.1 Upgrading mesh and cluster-tree networks

It is perfectly possible to deploy and upgrade a cluster-tree or mesh network, propagating the upgrade from node to node. For applications like smart energy, where the product life of devices may be 20 years or more, this may be a mandatory feature. However, the complexity of this upgrade process should not be underestimated.

Any upgrade process, and particularly any process that propagates across networks, needs to be tested to death. Never upgrade to a firmware version that has not been extensively tested, as the robustness of a subsequent upgrade will depend on what you have just deployed. Ensure that there is always one, and preferably two, levels of recoverability in every deployed node and try to ensure that there are multiple entry points to the network to restart or recover an upgrade process. With mesh networks, which employ end nodes and routers, make sure you understand the order in which nodes are upgraded. It may make a difference in being able to roll upgrades through the network. It is also wise to deploy a network-management tool that can reliably determine and report the status of each node within the network.

If any of these elements is missing, consider what the cost will be of sending service personnel out to replace or repair the entire network before you initiate an upgrade process. That should be part of the business plan for the product deployment.

Assuming that your upgrade works, never assume that it will work the next time. You've just changed the firmware in every node, so expect it to behave differently. Make sure you pay the same degree of attention to testing the process before each and every upgrade.

Best of all, try to design wireless networks, whatever level of complexity they may have, that do not require field upgrades.

8.7 References

[1] G2 Microsystems, www.g2microsystems.com.
[2] GainSpan, www.gainspan.com.
[3] Redpine Signals, www.redpinesignals.com.
[4] ZeroG Wireless, www.zerogwireless.com.
[5] Ozmo Devices, www.ozmodevices.com.
[6] MeshDynamics, www.meshdynamics.com.
[7] Strix Systems, www.strixsystems.com.
[8] SkyPilot, http://skypilot.trilliantinc.com.
[9] Tropos Networks, www.tropos.com.

9 Application development – performance

In the previous chapter we looked at some of the parameters that affect the choice of a wireless standard. This chapter explains how to get the best performance by tailoring it for the specific implementation. It explains the trade-offs that can be made for some of the common parameters in a wireless design. As before, many of the comments and techniques are valid across the range of standards.

9.1 Range and throughput

Invariably, the first question that is asked is, 'What is the range?' In Chapter 2, I looked at the fundamentals of range, which are essentially: transmit power, receive sensitivity and matching. In this chapter, I'll look at how to put them into practice and discuss the other key influence – the choice of antenna.

9.1.1 Power amplifiers and low noise amplifiers

The first thought of most designers coming to wireless is how to shout louder; in other words, how they can add additional amplification to boost the transmit power. A number of points should be borne in mind when doing this:

As the radio link is symmetrical (i.e., each radio needs to receive as well as transmit), increasing the output power only gives a real benefit if it is done at both ends, otherwise the second radio will not be able to transmit at a level that allows the first unit to hear whether or not its transmissions have been received. It is back to the issue of asymmetric link budgets. That can be helped by increasing the device's receive sensitivity with the addition of a low noise amplifier (LNA), as shown in Fig. 9.1.

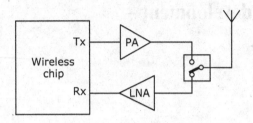

Figure 9.1 Power amplifier and low noise amplifier

This approach is particularly important where you don't own both ends of the link. As one of the reasons for choosing a wireless standard is to gain interoperability with other devices, this may often be the case. If you want to achieve an enhanced range with other connectable devices, simply increasing output power is unlikely to realise any significant benefit. You need to be able to listen efficiently as well as shout.

If you do own both ends of the link, then improving receive sensitivity is often a more cost-effective option than attempting to make the equivalent increase in transmit power.

Power consumption increases significantly with RF output power. There is not a simple relationship. Figure 9.2 provides an example of 2.4 GHz performance using a number of different power amplifiers. Even when not transmitting at full power, the quiescent current of a power amplifier can be significant. Again, this is an argument for looking at improving receive sensitivity, which has little impact on power consumption.

As power levels are raised, it can become increasingly difficult to contain the RF emissions within the regulatory limits (Fig. 9.3). This is particularly true where more complex modulations are employed in the coding schemes. As Fig. 9.3 illustrates, with OFDM coding at +20 dBm transmit power, the output can be very close to the maximum limits. If it is necessary to transmit at the maximum possible power, then multi-pole filters are likely to be needed, which add considerable cost and size to the product.

As power amplifiers themselves add distortion and non-linearities to the signal, they need to be chosen carefully to ensure that they are operating in the most linear portion of their range. This

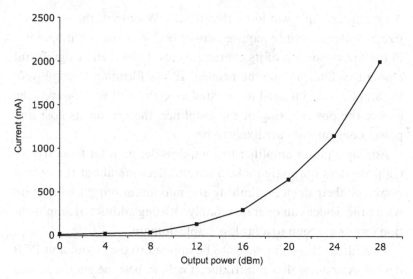

Figure 9.2 Typical PA power consumption

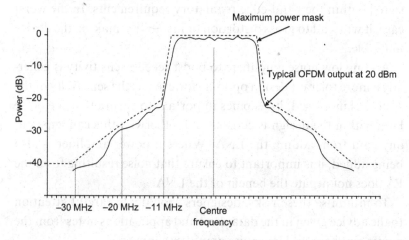

Figure 9.3 Spectral mask issues at high power

leads to a trade-off in the amount of filtering necessary to achieve band-edge compliance. The more linear they are, the fewer additional filters will be needed.

Power amplifiers generally have their most linear response when they are working well within their capabilities. As they approach their maximum rating, their performance becomes more non-linear.

A typical example would be that for a 1 W output, the minimum size of power amplifier required might be 2 W, to ensure it operates in the linear portion of its characteristics. Even then, a significant amount of filtering may be needed. If less filtration is employed, the amplifier might need to be rated as high as 10 W. However, the higher the power rating of the amplifier, the greater its cost and power consumption are likely to be.

Adding a power amplifier to a wireless design is far from trivial. Chip vendors normally make a careful decision about the output power of their devices, limiting the maximum output to a point where the device can operate reliably. Adding additional amplification moves a design to a far less stable domain.

To address this, care needs to be applied to the layout and PCB design. Any noise that is introduced will, at best, be amplified and added to the output signal, making it more difficult to keep the signal within the band-edge regulatory requirements. In the worst case, it will lead to gross distortion or instability, making the device unusable.

Adding low noise amplifiers to boost receive sensitivity is generally a more tolerant design option. However, as the sensitivity of the system is increased, it becomes important to ensure that the noise level within the design is constrained; otherwise this can wipe out any gain from adding the LNA. Where a power amplifier is also being added, it is important to ensure that noise resulting from the PA does not negate the benefit of the LNA.

To minimise these risks, designers should pay careful attention to the advice given in the data sheet and applications notes from the power amplifier and low noise amplifier suppliers.

Any experienced RF PCB designer will know that layout can be critical. One particular area that can be troublesome in 2.4 GHz designs is ensuring that the oscillator signals are kept clean. In most 2.4 GHz wireless chips, any digital noise introduced onto the oscillator will cause major performance issues. It typically couples into the phase-locked loops which generate the internal demodulation signals and adversely affects the receive sensitivity. Take great care over the tracking of oscillator signals and the placement of components. Where possible, PCB designers should use ground planes to isolate them from digital signals. Noise from poor layout can easily destroy any gains achieved by adding an LNA.

9.1.2 Power control

Some standards mandate power control for devices with high transmit powers, others do not. When implementing high-power designs it is useful to see if the topology and device capabilities allow power control and if so, to implement it.

The advantages of controlling output power are twofold. By running at a level below the absolute maximum, the power consumption savings will be significant. It also helps to reduce the overall level of noise going into the spectrum, which will make all of the radio links within an area more efficient. It is exactly analogous to a noisy room full of people – when there is a lull in conversation, the sound level drops noticeably, and, if you are lucky, remains at a lower level when conversations resume.

In topologies where there are large numbers of nodes that transmit asynchronously, then other considerations start to come into play. Where all of the devices are operating on the same frequency, it becomes increasingly likely that another device within range may be transmitting at a point where a node wants to transmit itself. This is more likely to be a problem where a standard is used that works at a fixed frequency. These often use a listen-before-transmit scheme, so the node will delay its transmission, resulting in an increase in latency and a lower throughput. In an extreme case, this can cause a frequency-agile network to try and change channel, which can result in the network becoming partially unavailable for a significant amount of time. The greater the ratio of transmission range vs average internode distance, the more severe the effect will be. When designing this sort of network there is a strong incentive to implement power control or to design the power output based on the anticipated deployments.

9.1.3 Filtering

When the need arises for a design to transmit at a power level close to the maximum allowed by the standard or the national regulatory bodies, attention must be paid to the emissions spectrum, especially at the band edges. The critical points are normally at the top edge of the spectrum for both Bluetooth and Wi-Fi, where there is a narrower guard band. The higher the data rate and, hence, the more complex the modulation, the greater the problem is likely to be. (ZigBee and Bluetooth low energy are unlikely to experience problems, as both use simpler modulation schemes, resulting in much better defined spectral output.)

At output powers above 18 dBm, both Bluetooth and Wi-Fi are likely to require additional filtering, for which the most appropriate candidates are monoblock ceramic filters or bulk acoustic wave (BAW) filters. Bulk acoustic wave filters are only just appearing on the market, and are limited in their range. For Wi-Fi, where the upper band edge varies from country to country, this presents a problem, as it may necessitate the use of a different filter for products shipping into different countries.

The comparative cost and complexity of using filters for wireless products is a good reason for limiting the maximum output power. Squeezing out the last few dB of potential transmit power can double the cost of a 2.4 GHz wireless design.

9.1.4 RF matching, tuning and PCB design

One of the commonest mistakes in wireless design is a failure to match the components within the RF section of the design. At 2.4 GHz the PCB acts as a component and needs to be designed as such.

Equally importantly, all components within the RF section need to be matched to each other, typically requiring a 50 Ω impedance. Failure to do this can result in some or all of the output signal being reflected back, with minimal amounts ever reaching the antenna.

Before starting work on a PCB design, talk to your PCB supplier to determine the material grades and the process and layer stack recommendations. The way in which the PCB layer stack is built up will affect its impedance and RF characteristic.

During layout, ensure that all track lengths are kept to a minimum and that component pads do not allow the component position to wander. Tracks must be designed to have a 50 Ω impedance, which is controlled by their width, length and what is on the other layers of the PCB. If antenna tracks need to be long, make sure that the length is not close to being a half wavelength.

Ground planes need to be regularly 'stitched' together using plenty of vias, otherwise they have a tendency to float. At RF frequencies, never assume that the potential of a plane is the same across its area. It is worth using buried vias to allow as solid a ground plane as possible, rather than having it eaten away by through-hole vias. Flood all via connection to ground and power planes to minimise the impedance of the plane and via connection. This can make a significant difference to performance.

In designs using a power amplifier, it may be useful to design a variety of locations for tuning capacitors and inductors on the power track for the PA. This will allow an experimental approach to be taken with the optimum placement for the components. It is a good illustration of the effect that component position can have in RF designs. It's not unusual to see a variation of several dB in output power just as a result of the position of the decoupling components.

Although not technically matching, the insertion loss of components used in the RF chain should be carefully considered. Every passive component will result in a loss of a small amount of signal. That is equally true for diodes, switches, connectors, cables, inductors, resistors and capacitors. Typically the value is only 0.5 dB to 1.5 dB per component or element, but these can easily add up to remove 5 dB or more from the power that eventually reaches the antenna. And the antenna can also add to the loss (as we'll see later). If a component is used in both the transmit and receive chain, insertion loss has a double hit on the overall link budget. There is no way that insertion loss can be removed, but a judicious choice of components and limiting the number of them can reduce it substantially.

Always consider the use of a screening cover for the RF section of a design. It is worth incorporating the mounting pads on a PCB design, even if the cover is not eventually used. Regulations in some countries may require one to be fitted, particularly if the PCB containing the radio is accessible by the user.

9.2 Choice of antenna

In the same way that lack of matching can ruin a good design, so can the wrong choice of antenna. There are many companies making antennae but often little thought is applied to selecting the most appropriate one.

Antennae have four main parameters, which are important to the short-range wireless designer:

9.2.1 Gain

All antennae have an effect on the signal, either increasing it or reducing it. This is referred to as the antenna gain, which is specified in dB. The gain is added to the link budget and, in the case of a shared antenna design, affects both the transmit and receive signals, so has a twofold effect on link budget. Positive figures for gain are normally associated with directional antennae, where the gain is concentrated in one direction at the expense of lowered sensitivity in other directions. At the other end of the range, negative gain is common in small ceramic or printed antennae, where the size of the element is considerably less than the wavelength.

9.2.2 Directionality

Some antennae are designed to be highly directional, particularly where a signal needs to be transmitted between two distant points. For short-range wireless, where devices are often mobile and the antenna orientation will be undefined, the preference is normally to choose a largely omnidirectional radiation pattern so that the orientation of devices does not have a great effect on range.

Omnidirectional is a fairly broad term when applied to antennae. Most omnidirectional antennae have a distribution pattern ranging from egg-shaped to doughnut, with a variation in gain of at least two-to-one across the sphere. Other antennae, such as printed-PCB antennae, can have much greater variation in their spatial gain distribution.

9.2.3 Construction (technology) and size

Although a fairly obvious parameter, antennae come in many different shapes and sizes. There's a fairly good correlation between size and performance – in general the bigger the antenna, the higher its gain. Highly directional antennae are normally larger, to accommodate the directional elements.

The construction of an antenna can affect its properties, particularly its frequency response. Small ceramic-patch antennae can offer good omnidirectional performance, but tend to be more expensive for a given gain than a larger external antenna. They also exhibit a band-pass characteristic, which may have a steep attenuation outside the frequency band. This can be used to advantage to lessen the need for band-pass filtering, but it does make these antennae more susceptible to antenna detuning.

9.2.4 Detuning

One of the least considered aspects of an antenna is its ability to be detuned by its surroundings.

Antennae are designed to have a centre frequency at the middle of their required frequency band. Depending on the antenna characteristic, its gain (generally measured as return loss) may not extend much past the frequency band (Fig. 9.4). These characteristics are normally quoted for the antenna when it is located at an optimum distance from an infinite ground plane.

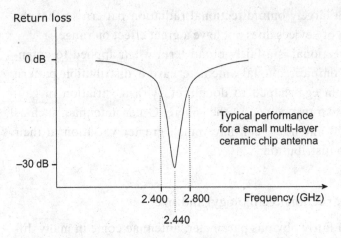

Figure 9.4 Antenna characteristics

In small, portable devices antennae rarely have anything resembling an infinite ground plane. For susceptible antennae, of which small ceramic ones are a good example, the lack of ground plane reference means that the centre frequency of the antenna can shift significantly depending on the mechanics around it, such as the plastic case, or the person holding or wearing the product (Fig. 9.5).

For body-worn devices, such as headsets, it's not unusual to see the centre frequency of an antenna move by 100 MHz as it is moved from a test bench to being worn on an ear. That can take the antenna gain completely outside the 2.4 GHz band, resulting in a drop in actual output power of several tens of dB. If you are using small antennae in portable devices, it may be necessary to choose ones with an offset centre frequency that 'detunes' back to the correct frequency in normal operation. If you do this, make sure that regulatory testing is done in conducted mode and not over the air.

The effect can easily be observed by attaching an antenna to a network analyser, feeding it at 2.4 GHz and then bringing a finger close to the antenna and watching the offset.

9.2.5 Polarisation and antenna radiation characteristics

One often-forgotten aspect of antennae is polarisation. I've touched on it above in talking about radiation patterns, but many types of

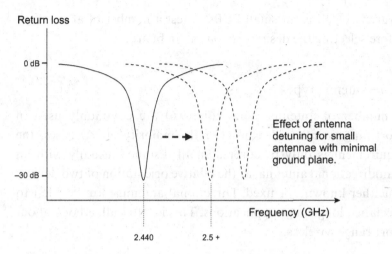

Figure 9.5 Antenna detuning

antenna are polarised as well. That means that if the transmitting and receiving antennae are orthogonal to each other, then very little signal will be received.

For small, mobile devices, polarised antennae present a real problem, as the relative orientation of the two antennae is likely to be changing constantly, resulting in what appears to be a random link budget. Unless you have a specific requirement and understand the issues that come with polarised antennae, it is best to avoid them.

9.2.6 Ground planes

The ground plane is an integral part of most antenna designs. When you read antenna data sheets, the performance data that you see will be for an optimised ground plane. In real designs, the antenna ground plane may well be different from this optimum, with the inevitable result that the performance is likely to be degraded. Even if the recommended ground plane is copied exactly, other nearby metalwork in the design may still have an effect on the radiation pattern.

For optimum performance, it is important to duplicate the recommended ground plane as closely as possible, as well as the antenna feed. Antenna manufacturers designed it the way it is for very good reasons. If you need to adjust it, then try making a variety of different small PCBs to test a number of alternatives before selecting the design for your main board.

9.2.7 Antenna types

A number of antenna types (Fig. 9.6) are commonly used in short-range wireless designs. For the majority of use cases, the requirement is only to cover a small distance, usually with an omnidirectional antenna, as the relative orientation of two devices is neither known nor fixed. Directional antennae are best left to specialist, long-range applications. This is, after all, a book about short-range wireless.

9.2.7.1 Printed (PCB and fractal antennae)

For the lowest-cost designs, an antenna pattern can be incorporated into the PCB tracking. PCB antennae tend to be susceptible to detuning and can exhibit directionality. Some of this can be overcome with more complex designs of printed fractal geometries, although a number of patents cover this latter technique. The low cost of a printed antenna needs be weighed up against the extra size of PCB needed for the tracking.

If the requirement is only for very short range, less than a few metres, then almost any small stub of PCB track may prove to be sufficient as an antenna.

9.2.7.2 Helical antennae and rubber ducks

Rubber-duck antennae are commonly used for access points, or wherever removable antennae are needed. They consist of a helical winding that is covered with a rubber or plastic housing. Common models provide useful gain of 2 dB to 7 dB and have maximum sensitivity in the plane at right angles to their length. Small helical antennae are becoming available in surface-mount format for direct placement on PCBs.

Most rubber duck (and other removable) antennae come with an SMA connector with a reverse thread. The history of the reverse

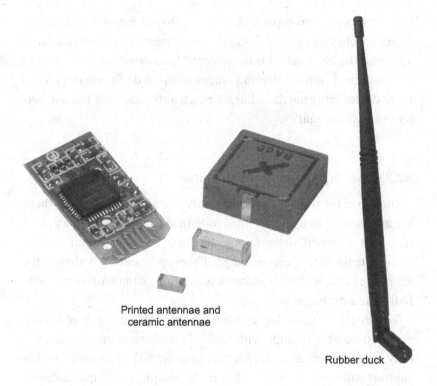

Printed antennae and
ceramic antennae

Rubber duck

Figure 9.6 Examples of antennae

thread is interesting. The FCC mandated it on the grounds that at
the time that radio devices were becoming popular, most antennae had a normal thread. As it is technically illegal to change an
antenna on a piece of certified equipment, the FCC decreed that
the chassis connector should have a reverse thread, so that standard, commercial antennae would not be compatible. The predict-
able result is that it is now almost impossible to obtain an antenna
with a standard thread – reverse threads have become the norm. (It
is still technically illegal for users to change the antenna on most
RF devices.)

9.2.7.3 Ceramic antennae
The growth in small, portable wireless products has seen considerable innovation in ceramic multilayer chip antennae. These range

from extremely small packages of less than 0.5 mm × 1.0 mm to larger patches up to 25 mm square. The larger the size of the patch antenna, the less likely it is be susceptible to detuning.

The gain of patch antennae ranges from −5 db for small patches to +2 dB for large patches. Larger patch antennae tend to have better omnidirectionality.

9.2.8 Diversity and multiple antennae

Although most wireless products only use a single antenna, there are advantages in using multiple antennae. By using receive antennae with different orientations, the best received signal can be chosen, helping to increase range. This is particularly valuable for short-range radio, where indoor installations can result in variable fading and propagation.

To benefit from multiple antennae, the RF circuitry of a chip needs to be able to cope with multiple feeds. For many years the only common implementation was in some 802.11 chipsets, which support antenna diversity. This is the simplest multiple antenna scheme, where one antenna is always used to transmit, but two antennae are used for the receiver. The receiver accepts the stronger of the two incoming signals and uses that.

A more complex variant is spatial diversity, where the two incoming feeds are combined, with additional processing that generates a better signal than would be available from either independent feed.

802.11n heralded the arrival of more advanced antenna techniques with the introduction of MIMO, where multiple transmitters and receivers are connected to separate antennae to provide a degree of beamforming. This allows the radio link to be made more directional, giving a useful increase in range over an omnidirectional antenna. Although these techniques could be applied to any radio standard, in practice the cost is prohibitive unless it is incorporated in both the standard and the resulting silicon.

9.2.9 One last point on antennae

One final comment on antennae is not to enclose them in a metal box, or use a metallised or conductive plastic housing. It is surprising how some industrial designers still think that wireless will work inside a Faraday cage.

9.3 Coexistence

It is not unusual for a company to produce a wireless design that works perfectly in their building, only to find themselves deluged with support calls when the performance elsewhere is underwhelming or non-existent. In many cases, that is the result of interference. All of the 2.4 GHz standards share the same spectrum, which, depending on where you are in the world, means a variety of proprietary radios, microwave ovens, cordless handsets and industrial equipment. If there are enough transmitters operating, everything will grind to a halt.

The bad news is that the problem is likely to be least severe on the day you ship your first product and then become progressively worse as time goes on, as more and more products are shipped using this band. In the course of time, matters may improve as standards employ better interference-mitigation strategies and heavy users move to higher frequency bands, such as 5.1 GHz. But there is no guarantee, so designers need to do two things:

• Test their products thoroughly in a variety of different locations,
• Look at ways to build in interference-mitigation strategies.

9.3.1 Interference mitigation

It is surprising how little the different standards have done in terms of accepting the issue of shared spectrum and developing strategies to cope with them. Although some work has been performed within the IEEE 802 groups, it has yet to be incorporated into the

Wi-Fi or ZigBee standards. Bluetooth has adopted a frequency-hopping strategy from the start and enhanced this with an adaptive hopping scheme relatively early on in its life. This has a dual benefit of moving to frequencies within the band that are clear, and also avoid treading on the toes of fixed transmitters located nearby. However, the minimum number of channels that it uses is limited to 20, largely as a result of certain national regulations, which try to ensure that the transmitted energy is spread throughout the spectrum. That means that if the entire spectrum is occupied by other wireless transmitters that are static or slow hoppers, then Bluetooth can be a severe interferer.

When embarking on Wi-Fi or ZigBee designs, it is worth taking advantage of the interference mitigation schemes that they provide. Both of these standards normally transmit on a fixed channel. Most products also ship with the same default channel as standard. If your product allows it, there is an obvious advantage to choosing a different channel as your default. This is particularly useful for ZigBee. Both standards also provide the ability to choose the channel dynamically after a scan of the area. Again, this is a sensible approach as part of the automatic set-up for a device. The most recent ZigBee PRO standard provides the option for a network suffering from interference to choose a new channel to operate on. It should be considered as an essential feature of any product sold commercially.

Good design of receivers can help and it is worth comparing the performance of different chips, as the blocking performance varies considerably between manufacturers and even between different generations of chips from a single manufacturer. It is one of the parameters that is not on data sheets and needs to be assessed by physical testing. Where interfering devices are located within range of each other, this can make the difference between a product working and not working.

There is not much else that can be done to combat interference from other uncoordinated emitters within the same band. Where possible, try to limit the output power to the minimum that is required for reliable operation. It may not help your product, but it will help to reduce interference in others. With luck they will adopt the same strategy.

9.3.2 Colocation

Colocation is the special case where two different radios are located within the same device. The most common examples are Bluetooth and Wi-Fi, but other combinations exist, including ones which include proprietary radios operating in the same band. As well as the normal issue of interference when they are operating asynchronously, colocation introduces the new and potentially more concerning issue of front-end overload.

Front-end overload occurs when a transmitter is adjacent to another receiver, resulting in a much higher input signal than the receiver was designed for. When this occurs, the receiver can saturate and may take some considerable time to recover so that it misses information it should be receiving.

Where the two radios are located sufficiently close within a single piece of equipment, it is important that they share information about when each is transmitting or about to transmit, so that they can coordinate with each other. This is not a simple task. To be able to react quickly enough the radios typically need to be able to communicate at the baseband level.

Some chipsets provide a variety of signals that can be used to implement basic colocation schemes, normally by delaying transmission and warning the other device when transmission is about to occur. There are no standardised colocation schemes – different manufacturers have taken different approaches. There is limited interoperability within these approaches and the effect of combining chips from different vendors can be unpredictable.

If you are planning to colocate Bluetooth and Wi-Fi in a design there is a new generation of combined chipsets designed for the mobile phone industry that include both radios. Many of these chips also include Bluetooth low energy support and the new version 3.0

Bluetooth standard that specifies the use of 802.11 over Bluetooth. As both radios are being provided from the one manufacturer, they normally incorporate proprietary signalling to optimise the colocation. If your design calls for this combination of radio standards, then these are likely to provide the best colocation performance.

9.4 Power consumption

One area where the different standards have virtually nothing in common is the way they treat power consumption. Most started life with some very basic concepts that have progressively evolved as they have matured. Today, the latest versions of each of the standards provide a variety of options that help to reduce power. Sadly, a large percentage of designs entering the marketplace continue to use older variants of the specifications or do not implement the most effective power-management strategies. One of the reasons that this happens is that power management is not mandated. Instead the power conservation techniques exist as a toolset within the standard, which can be used to increase the battery life of products. That often requires a detailed understanding of the specification.

One of the problems with interoperable standards is that power management often depends on it being implemented correctly at both ends of the link. Where it is not, or where different versions of the standard are used, as may be the case with a Wi-Fi access point and a laptop, the actual power handling tends to default to the lowest common denominator. It is not unusual to find that features that have been incorporated into a product do not get used when it is deployed in the real world, because both ends need to support them. For this reason, designers may want to consider how a product should work if it connects to another, allegedly interoperable, one that does not support its power-management modes. In cases where this might impinge negatively on battery life, it may be judicious to enter another operational mode, or inform the user of the problem if there is a user interface available.

It is beyond the scope of this book to go into fine detail of the different techniques in each of the standards. Other volumes, which dedicate themselves to each, or the standards and white papers from the organisations themselves, are the place to go for the latest information.[1, 2] What we will cover are some basics which are relevant for many designs.

If power consumption is an important parameter for your design, start by investigating the most recent chipsets and the most recent versions of the wireless standard you intend to use. Chipset vendors and standards groups both work hard to reduce power consumption with each new release, as they are under enormous pressure from their major customers to do so. Hence implementations with the same functionality can expect to see incremental and often useful reductions in overall current, just by moving to a new chip or stack.

9.4.1 Duty cycle

Chapter 2 explained the basic message of power consumption, which is to concentrate on staying asleep as much of the time as possible. Battery-powered designs need to start by looking to see how they can minimise the amount of time that they need to be awake to communicate and aim to go to a low power mode for the rest of the time.

That ideally means turning the radio off altogether, as the receiver can take as much current as the transmitter. Unlike wired systems, be they telephones or Ethernet ports, there is no ability for low-level monitoring of a signal and then waking up in the way that modems or LAN cards can wake on an external signal. Therefore, devices need wake up strategies to see whether they are being requested to respond to a controlling device, or else to send data based on a local event.

This wake time is normally set by the controlling device. Low-power units should be told the approximate time they need to wake, set their low-power clocks and then wake up in time to stabilise and receive any incoming transmissions. They will normally have a window set, after which they will return to sleep if they have not received any data. Alternatively, they may work by assuming the recipient device is always listening and wake up asynchronously to send data or to poll to see if they have data or instructions waiting to be retrieved.

No one system is intrinsically better. The optimum scheme will depend on the volume of data to be transferred, duty cycle and the abilities of the master device to direct its slaves. What is important is to minimise the time a device is awake without being in an active communication, as that is what uses up the battery.

Where response time is important, as with human interface devices (mice and keyboards) or control circuits (which range from robot arms to process control valves) low latency response will determine what connection cycles and sleep modes are possible. Although this imposes a compromise, wireless standards that address these applications have developed specific approaches, such as the sniff sub-rating in Bluetooth version 2.1 and above. ZigBee and Bluetooth low energy address this problem with very responsive protocols, allowing them to sleep for most of their life, but respond rapidly.

It is important to remember that wireless devices may do more than just wake up when they need to send data. A good case is alarms. They may only expect to operate once or twice a year, yet they may need to report that they are operational and indicate their battery status on a regular basis, which may be as often as every few seconds. It's not uncommon to see power requirements worked out that totally miss these management transmissions, even though they are responsible for 99% of the total wireless power budget. Another item often forgotten is the power consumption for the initial configuration, where a device can remain powered for several minutes, wiping out most of the battery life before it begins operation. Remember to take into account everything that a radio does and everything that you have allowed a user to do to affect its battery life, otherwise you may get an unwelcome surprise when it is deployed.

9.4.2 Sleep modes

When a radio goes to sleep, it normally has a choice of a number of different possible sleep modes. These may be specified within the standard, be proprietary to the chip manufacturer or be a combination of both.

The reason that multiple modes exist is that different applications need the chip to be able to respond to a wake-up in different ways. Many chips save cost by using an external memory or host processor to download their firmware code into RAM each time they power up. (It is cheaper to use commodity external storage rather than integrating the same volume of flash within the main chip.) For fastest start-up from sleep, chips do not want to reload this, so there is almost always a sleep mode that keeps the RAM powered at a sufficient level to allow it to retain its content. This is complemented with deeper sleep modes that will progressively turn off more circuitry to the point where the device is completely off, with the possible exception of a low-accuracy 32 kHz clock and a monitor that checks for a wake-up signal on one or more designated ports. In these, the start-up time may need to be greater. It is another compromise.

Designers need to look at the balance of consumption and response time for the different sleep modes and weigh up the conflicting requirements of response time and overall battery life.

9.4.3 Functional circuitry

It is important to ensure that the wireless element of a design is optimised to use the minimal possible power. In many cases, though, it may not be the major consumer of power within a device. It is not uncommon to see many hours spent optimising a radio, only to discover that it is responsible for just a few per cent of overall current consumption.

Every application is different, so there are no hard and fast rules. Designers should make sure that, in sleep mode, the rest of the circuitry can go into and come out of sleep mode at a similar rate to the wireless section, so that neither is burning power waiting for the other. Whereas most wireless chips are designed for very good sleep-mode performance, the same is not true of many other electronic components, so these should be chosen with as much care for sleep current consumption as the radio itself.

It is not uncommon for the major source of current consumption over a device's lifetime to be the sleep mode current, particularly where very low duty cycles are involved. Always make sure that an audit of power consumption includes every part of the circuit and all of the different operating states of each of those parts.

9.5 Topology effects

It can be difficult to work out the power consumption parameters for two devices that provide a simple cable replacement connection. As topologies become more complex and devices need to fit into a more complex connection schedule, they may end up needing to be awake for longer, introducing another factor into power-consumption calculations.

This becomes more important with the expansion into star or mesh networks, particularly if they allow low-power devices to take part in connections with several masters or routers, which can control their low-power modes. In this case, multiple connections may enforce different duty cycles and sleep modes. This leads to the situation where it's possible for a device to get caught in a state where it needs to be on all of the time to cope with conflicting demands. In these more complex topologies, it is important to understand who controls each aspect of the network's power consumption. A similar case can occur with Wi-Fi, as a device roams between access points. If the different access points support different sleep modes, the client device may be forced into a less optimal power-management mode than the designer intended.

Always assume that additional complexity may increase power consumption and try to model the worst case.

9.6 Ultra-low power and energy harvesting

The holy grail of low-power wireless is to remove the need for a power supply and rely on energy harvesting. The better-performing chips implementing 802.15.4 and Bluetooth low energy have just about reached this point, where it is possible to use thermal or vibrational power to run them.

Other approaches, such as the ultra-low power chips from companies like EnOcean have highly optimised radio modes, allowing them to operate from the energy generated by piezoelectric elements in switches.

The area of self-powered wireless sensors will start to take off in the coming years. Today it is just becoming possible, but needs great attention to component choice, power management and connection duty cycles. With one further generation of chips and the anticipated improvement in sleep current, the market is likely to grow substantially.

9.7 Temperature

Although none of the wireless standards specify a temperature range, when you survey the chips and modules on the market, it quickly becomes apparent that almost all are designed for consumer applications, with temperature ranges of around 0 °C to 50 °C. A few vendors have offerings that are targeted at the automotive market, which extend this to −40 °C and as high as 125 °C and some modules are available for the industrial range of −40 °C to +80 °C.

Do not make the mistake of assuming that a wireless chip rated for 0 °C to 50 °C will work over a wider range. Many of the components of the chip have temperature dependencies and these may combine to seriously compromise performance or even stop them working. Almost all of the market for short-range wireless is in consumer devices, which are rarely operated outside the human comfort range of 10 °C to 40 °C. Taking a design beyond this can be a surprisingly complex proposition.

If your product needs to operate over an extended temperature range, the first task is to ensure that all of the components are rated to operate over that range. Although that may sound a trivial exercise, components such as crystal oscillators are designed for consumer applications and their accuracy is rarely specified at extreme ranges. It is vital to ensure that every component is appropriately rated.

Be aware that if a critical component within a reference design, such as the crystal or balun, is changed to achieve the correct temperature specification, then it is likely that the product will need to be requalified, or at least taken through a delta approval. This means that extending an operational temperature range can be a costly exercise.

9.7.1 Working below 0 °C

A number of separate effects can come into play as wireless devices are operated at lower temperatures. The most severe is for the receiver to 'go deaf', which, when it happens, normally kicks in around −25 °C. It is important to test a good range of chips from each supplier to ensure that you design with a representative sample, or else obtain a selection of edge samples to test. This particular problem can be alleviated on high-power designs by including a small heater on the PCB to raise the chip temperature at start-up. Once these are running, the die temperature will normally rise to a point where it will work normally. However, this is an extreme route and it is normally safer to look for an alternative chip. In the case of low-power products, incorporating a heater will kill the battery life, so the only alternative is to choose another chipset.

The other common issue is drift in the oscillator circuits, either within the chip or in the external crystal. Although these will probably not stop the device working correctly, there is a strong possibility that they may take it outside the spectral requirements so that it fails regulatory testing. If the cause is in the external crystal, it can normally be solved with a higher-specification crystal or, in the extreme case, with a temperature-controlled oscillator, although that will add to the cost and the current consumption. If the effect is in the power-amplifier stage, then it may be possible to correct this with temperature-compensation circuitry. If not, it generally requires the use of an alternative chipset.

9.7.2 Working above 50 °C

At higher temperatures, the same issue of oscillator drift exists. Here temperature-controlled oscillators are unlikely to be of much help.

The more severe problem, particularly with higher-power chips, is that of overheating. This can be contained if it is possible to limit the output power at higher temperatures, either by using an active thermal compensation circuit, or by measuring board temperature and using software to reduce the output power to a sustainable level.

For single-chip solutions running high-duty cycles or streaming at output powers of 10 mW or more, it may be necessary to provide heat sinking for the main wireless chips. If higher powers are required at elevated temperatures, then some expensive cooling is likely to be needed.

9.8 References

[1] Bluetooth Special Interest Group, *Bluetooth Sniff and Sniff-Subrating Modes Whitepaper*. www.bluetooth.org/DocMan/handlers/DownloadDoc.ashx?doc_id=125640.

[2] Wi-Fi Alliance, *Support for Advanced Power Save for Mobile and Portable Devices in Wi-Fi Networks*. www.wi-fi.org/white_papers/whitepaper-120505-wmmpowersave.

10 Practical considerations – production, certification and IP

Many designers rush into wireless without any knowledge or consideration of the practical issues they will face in manufacturing and selling a wireless product. Wireless introduces a number of requirements over and above those of normal electronics design. These need to be understood if manufacturers wish to place their products on the market and conform to legal requirements.

This chapter highlights these areas, so that a designer can assess the most practical route when embarking on a wireless design. If they are ignored, (as they frequently are), the resulting cost in putting things right after the event can be greater than the cost of the rest of the design effort. In the worst case, a national regulator can stop shipment of products within its country.

10.1 Regulatory approval

To the best of my knowledge, it is legal to sell a cable anywhere in the world. Plugging in a cable doesn't generate any significant amount of electromagnetic radiation that could interfere with other products. Replace that cable with a radio transmitter and everything changes.

Although we are talking about radios that work in the unlicensed ISM bands, that does not grant designers a right of laissez-faire. Products still need to adhere to strict rules and manufacturers must be able to prove that they meet them. These rules exist to try to ensure open access to anyone who wants to use that spectrum, minimising the possibility and severity of interference and to prevent any single product from monopolising too much of the spectrum. Although these regulations vary from country to country, they generally impose maximum levels for transmit power and limit the overall amount of power which can be pumped into the spectrum.

During the late 1990s, the Bluetooth SIG lobbied regulators around the world to try and get an even regulatory playing field, with the same amount of spectrum for the 2.4 GHz band and the same requirements on RF transmission. These efforts were largely successful, but there still remain some discrepancies.

In terms of the way that a radio uses the spectrum, i.e., its modulation and channel width, the parameters are defined by each standard, which is based on the regulatory landscape. Similarly, the overall transmit power is normally limited by the standard. Having said which, there are some notable exceptions.

Of these, the key ones to be aware of are differences in the number of available channels for radios and limitations in the maximum output power. What this means for manufacturers is that if they are shipping these products into countries where there are different requirements, they need to preset the appropriate levels at manufacture and produce a number of country-specific variants.

These regulations do change. Manufacturers and designers should always check with their test house, or the national regulatory body, to determine the most up-to-date requirements for countries in which they intend to sell their products.

Before a product can be shipped, it needs to be tested to ensure that it complies with the national requirements. This can be done in-house if suitable expertise and equipment are available, but is more normally performed by an external test house. The key sets of requirements are defined by:

USA: FCC [1] Part 15.247,
Canada: ICES-0003,[2] RS-210A8,
Europe: CE,[3] ETSI 300 428 RF, ETSI 301 489–1,ETSI 301 489–17 and EN60950,[4]
Japan: TELEC,[5]
China: CNCA China Compulsory Certification (CCC).

Other countries have their own requirements, but they are generally similar to the CE and FCC requirements. In some cases, they will accept test reports that have been performed for CE or FCC.

If products are to be sold within Europe, manufacturers do not need to submit evidence of compliance, but must maintain a technical folder containing test results showing compliance. If products are being sold in the USA, then test evidence must be submitted to the FCC and approval obtained before products can be shipped. Products need to reference the FCC approval number for that specific piece or family of equipment.

10.1.1 Modular approval

Some regulatory authorities, notably the FCC, allow modules to be submitted for approval. These then carry a modular approval, which can be cited by a manufacturer, removing or reducing the need for further RF testing. This still imposes strict requirements, which dictate whether or not the final product needs retesting. Most commonly these cover:

- The antenna. Modular approvals specify the antenna used for that specific approval. If any other antenna is employed, then an incremental retest will be required. If the gain of the antenna is greater than that used in the modular approval test, a full retest may be needed.
- There's a common belief that if a module contains an RF connector, then any antenna can be used with it. This is fallacious. Using any antenna other than the specified one will void the approval. However, it is normally possible to persuade a module manufacturer to request a delta approval to cover the use of an alternative antenna. This is cheaper than performing a full approval.
- Multiple modules. If more than one different radio is employed within a product, then most regulatory bodies will not accept a combination of individual approval certificates, but will demand a retest to check that the combined effect of the multiple radios still lies within the national limits. If the radios talk to each other to synchronise their transmission in any way, there will almost always be a requirement to retest.

• Additional amplifiers. Modular approval does not permit any change to the radio output. If amplifiers or other circuitry are added between the output and the antenna, a full retest is required.

10.1.2 Other considerations

It is important to note that these approvals cover the *whole* equipment – not just the wireless portion. If pre-approved modules are used, the module vendor's approvals can be submitted as evidence of compliance for the radio part of a product, but a manufacturer will still need to complete an overall approval for its product before it can be legally placed on the market.

10.1.3 The Radio and Telecommunications Terminal Equipment directive (R&TTE)

Within most of Europe, 2.4 GHz products with an output power less than 10 mW (+10 dBm) can be sold freely, subject only to meeting the relevant CE requirements. Once the output power exceeds 10 mW, the situation changes and products have to be notified to the regulatory authority of each country in accordance with the Radio and Telecommunications Terminal Equipment directive (R&TTE).[6] At some point in the future, this requirement is likely to change, but as long as countries like France have different requirements, resulting in a non-harmonised market, it is a process to which vendors need to conform.

Anyone designing Wi-Fi products, which transmit above 10 dBm in their normal operational mode, needs to go through this procedure. Most Bluetooth and ZigBee products transmit between 0 dBm and 6 dBm; hence, they do not need R&TTE notification.

To submit a product, manufacturers need to complete the appropriate forms and submit them. There is an automated process, available on the web at the European Commission Enterprise e-services Portal,[7] which simplifies the procedure, and which covers most participating countries. Submissions must be made three months prior to the first intended sale. This timeline should, therefore, be planned and communicated to sales and marketing departments. Products can be shipped after that three-month period, or sooner if acknowledged by an individual country. Notification is currently free, except for Switzerland, which charges a nominal fee.

10.2 Specific absorption rate – SAR

For products that are worn, or which come into close contact with the skin, additional testing needs to be performed to measure the specific absorption rate (SAR).[8] In general, this is necessary where the product will regularly be used within 20 cm of the human body.

10.3 Medical, automotive and aviation

These three markets have a wealth of approvals and regulations. Wireless is a relatively new territory for all of them, so even where there is an industry approval, the actual deployment of devices may still vary on a site-by-site basis. Current examples are the hospital-by-hospital guidelines on the use of Wi-Fi, Bluetooth and ZigBee. Whereas some hospitals ban all three, others have full wireless infrastructure coverage. Aviation is similar, with the use of Bluetooth and Wi-Fi allowed by some airlines and banned by others.

In time, these guidelines are likely to coalesce and allow universal wireless use. In the short term, it means that products may be allowed in one location but not in another, even within the same town. Where the wireless link does not need to be permanently on, it makes sense to add a prominent 'wireless off' feature (also known as 'flight-safe mode') to a product to reassure people that it can be used safely.

In the medical market, the Continua Health Alliance [9] is acting as an industry body to try and standardise both the data formats and protocols used in medical devices, and to bring a degree of consistency to regulations governing its use. They currently support Bluetooth BR/EDR in their guidelines and will add ZigBee and Bluetooth low energy to their next releases.

For medical devices, the major regulatory body is the USA's Federal Drugs Administration – the FDA. Although it only covers medical products sold within the USA, it has some of the toughest and most wide-reaching powers and is the primary regulatory barrier. In most cases, if a product can achieve FDA certification, it will pass the requirements of any other country.

The FDA publishes guidelines for manufacturers, which helps to guide them through the process of certification.[10] However, much of the FDA's experience is based on traditional products with displays or cabled links. Wireless connectivity and devices that feed data directly to an electronic health record (EHR) are new territory. A good overview of the issues this raises has been provided by Bradley Merrill Thompson.[11]

10.4 Export controls

Governments around the world are extremely sensitive about technology becoming available to countries which they perceive as a military or political threat. As the level of technology within consumer products has advanced, this has resulted in lists of prohibited or controlled technologies, which require official consent if they are to be exported. Whilst many countries operate a 'dual-use' policy, whereby technology embedded in consumer products is regarded as inaccessible and safe to export, there is often still a need to register the intent to export with the appropriate authority.

Unfortunately for the wireless community, encryption remains high on the list of controlled technologies. As the various wireless standards have introduced higher and higher levels of encryption into their core standards to safeguard the wireless link, they have strayed into the area of controlled technology. For most countries the level of control kicks in for encryption keys greater than 56 bits, which includes the stronger encryption recommended by the standards covered within this book.

Unless a lower level of security is employed, which may be appropriate for a device like a headset or a light switch, manufacturers will need to obtain an export license for their products. In the first instance, the national bodies dealing with this should be approached to discover the procedure and appropriate level of compliance. Even when a product is only intended to be shipped within one country, it is still worth checking the requirements, as unintended exports by a customer could still be considered a contravention, for which the manufacturer may be held responsible.

Care should be taken that downloadable software or firmware upgrades do not infringe export control regulations. An example is where a customer buys a product in a country where the encryption is limited to 56 bits, but is then able to download and install an upgrade from a website in another country that converts it to 128 bit encryption. In the eyes of most governments, a company that designed a product that allowed this to happen would be contravening their export controls.

When applying for export licences, it is important to consider all countries that a product, and the controlled components within it, will touch. That is particularly important where a product may be designed in one country, manufactured in a second country and shipped to a third.

Export control legislation also applies to the transfer of design and manufacturing data, so advice should be sought if design teams are spread around the world. Most government departments have yet to come to terms with the reality that design information, such as encryption software, may be distributed across many different sites within a single company's design department. If in doubt, seek the advice of the government departments in each country concerned and, where possible, get written confirmation.

10.5 Standards-based approvals and IP licences

All the major established standards discussed in this book run their own qualification programs. Although these are often seen as just an exercise in raising cash, they do confer a number of benefits to the manufacturer.

The first, most obvious, benefit is that they help to ensure that the product correctly implements the standard. This in turns gives confidence that they are interoperable with products from other manufacturers that have been similarly qualified.

The more important, and less understood, benefit is that the qualification gives the manufacturer a licence to utilise and ship the intellectual property (IP) contained within the standard. Depending on the particular standards body, this may or may not involve additional payment of license fees. It also gives the manufacturer the right to use the standard name and logo when marketing the product.

It is very important to understand what this means. If a wireless product is not qualified, then it does not have a right to use the intellectual property that the standards organisation and its members have contributed to the standard. That gives any of those IP holders – either the standards body itself, or any of the members, the right to take legal action against the infringing company.

The existence of a standard does not give a company a free right to exploit it, nor automatic access to the IP within it. To use it you must agree to the terms of the body that owns the standard.

Each of the different standards bodies treats this slightly differently, but there is a general agreement of principle. The starting point is that if you wish to use the name of the organisation to describe your product, i.e., Bluetooth, Wi-Fi or ZigBee, you need to sign up to the legal agreements of that organisation. Each has trademarked its name, and only gives you a right to use it when you sign its trademark agreement.

Signing up to the agreement places additional constraints on how you use the intellectual property within the standard. You need to confirm that you will adhere to the qualification program established by that body. You may also have to agree that you will not modify the product such that it works outside the bounds of the standard. In return, you are granted the rights to use the name and trademarks in accordance with the published procedures for each, and are given the rights to all of the IP within the standard.

There are some minor details to be aware of. One is the extent of the IP coverage. The IP rights are those included in the published standard. Different standards bodies impose different conditions on their members. As a general rule, any company contributing to a standard, agrees to make any IP that it owns, and which is incorporated within the standard, available to all users. This may be on a RAND (reasonable and non-discriminatory) basis, where the IP owners have the right to charge a reasonable license fee (although they rarely do), or on a RANDZ basis, where the use of IP is free from any form of license fee.

Before a standard is published, the standards bodies perform an IP review to ascertain whether any part of the standard infringes patents held by non-members (remember that when you sign up to a standard you agree to 'donate' your relevant IP). Patent searches can never give 100% coverage and there is always the possibility that something may emerge at a later date, but it gives a level of assurance that using the standard will not violate any patents. Obviously, the larger the membership of the standards organisation, the more IP will be covered by the member agreement, and the level of assurance increases.

Companies need to pay particular attention to the level of coverage of the standard. Bluetooth covers the entire wireless system, from radio to application interface. ZigBee and Wi-Fi do not – they both employ radios that have emerged from IEEE standards groups. These had far fewer companies involved in their development, so there is a greater risk of patents existing elsewhere that could be employed against manufacturers. A recent case has occurred with 802.11g, where a research institute has sued a number of Wi-Fi companies for infringement of core OFDM coding patents,[12] Manufacturers should assess the risk when using a standard and

Table 10.1 *Licensing requirements of short-range standards*

	Bluetooth	Wi-Fi	ZigBee	Bluetooth low energy
License	RANDZ	RAND	RAND	RANDZ
Annual membership[a]	Free	$5k/$15k[b]	$3.5k[c]	Free
Ownership of MAC/ PHY	Bluetooth	IEEE (802.11)	IEEE (802.15.4)	Bluetooth

[a] Minimum level of membership fee for use of trademark
[b] To certify a Wi-Fi product, the minimum membership level is regular
[c] Non-commercial users may use the ZigBee standard without payment

decide whether or not they should budget for the possibility of future license fees.

Table 10.1 indicates the current status of different standards and their qualification requirements. These change with time, so consult the appropriate bodies for the latest status.

Qualifications are performed by most of the test houses that conduct national regulatory certifications. In some cases there may be an overlap of regulatory and qualification tests, but most test houses will permit previous tests to be accepted. It is often cost-effective to have all testing performed in the same test house. Some qualification tests may be performed in-house, using certified software or test equipment, if the appropriate level of expertise is available.

Unlike regulatory testing, standards-based approvals are global. In practice there are a few countries that may impose limitations, such as France (which limits output power). But in these cases this is covered by the national regulations. Standards approvals do not negate the need for national regulatory approval.

10.5.1 Standards approval hierarchies

Most standards accept that it does not make sense to reapprove everything from scratch, particularly where a company may make several variants of a product. Therefore, they have a number of schemes that allow incremental approval. These may help designers to make the choice of using a module, a reference circuit or a chip-on-board design.

Most standards allow modules to be pre-qualified, so that using them does not require requalification. There are some subtle nuances to these rules. For Bluetooth, if an additional profile is added to a module, over and above what has been approved, it needs to go through a subsidiary qualification. ZigBee does not approve modules with full profile support, rather than ZigBee compliant platforms, so there is a similar fee for testing the final profile. Wi-Fi permits the use of approved modules, but places rules on the dependent platforms within which they can be used. Bluetooth low energy allows new GATT-based server profiles to be added at no cost, albeit they require verification of each one. If in doubt, check with a test house or standards organisation.

According to the letter of the qualification programs, any change to a qualified product, whether of a critical component, a new revision of software, or the addition of a new profile, will require a retest to ensure that compliance is still maintained. That is also true for software upgrades that are distributed to customers.

If you are designing a range of products using a wireless standard, it is worthwhile considering at an early stage how much of the wireless element can be standardised between them and how much can be shared within the wireless approvals. Clever reuse can have a significant effect in reducing the overall approval costs.

All of the standards bodies run regular interoperability test workshops or unplugfests, where manufacturers are able to test their products in a confidential environment with those from other companies. These can be immensely useful in ensuring that a product is interoperable.

10.5.2 Specific requirements

10.5.2.1 Bluetooth

The current version of the Bluetooth qualification program requires all Bluetooth products to have an end product listing (EPL). In most cases, this implies that the product contains a fully working Bluetooth radio and stack and at least one profile. If a manufacturer uses a module that has an end product listing and adds no additional profile functionality, then no further Bluetooth testing is required. The module's test reference number must be included within the product documentation and a free EPL must be completed on the Bluetooth SIG's qualification site.

The same applies if a reference design is used and evidence is available that no modifications have been made. However, manufacturers should be aware that any change of critical component from the supplier used in the reference design may invalidate the qualification. That will include any components in the RF section of the design, power-supply components or components associated with the crystal oscillator.

An exception is made to the end product requirement for Bluetooth products that provide an interface at the HCI level. These will typically be USB adaptors, or upper-layer host stacks, which are incorporated into products like PCs. In these cases, each product can be qualified as a Bluetooth subsystem. These can be shipped independently of each other and customers can expect the combination to work as an interoperable Bluetooth product.

Bluetooth cooperates with CTIA on their Bluetooth compatibility certification program,[13] which tests interoperability of handsfree products designed for used within cars.

There is one very limited exception to Bluetooth approval; products sold exclusively as test or development tools used to design or manufacture Bluetooth products do not need to be qualified. Normally these are specialised products that are sold to development engineers within the Bluetooth community. Only a few tens of products out of the tens of thousands of Bluetooth products on the market fall into this category.

Bluetooth qualification becomes slightly more complex with products incorporating higher-speed AMPs. At the moment this covers Bluetooth version 3.0, which allows the use of an 802.11 radio to provide a high speed ad hoc pipe. Both radios need to be qualified against the national regulatory requirements. Depending on the specific country (and sometimes also the philosophy of an individual test house), these may be considered to be independent or concurrent radios. As more of these products are tested, the national test requirements will probably become clearer, but designers implementing this mode should consult test houses in the target markets for clarification at the start of the design process.

It should also be appreciated that the 802.11 modes used by Bluetooth are not the same as the ones used within the Wi-Fi standard. Therefore, companies that want to use the 802.11 portion of the design in both modes will need to take the product separately through both Wi-Fi and Bluetooth certification.

10.5.2.2 Wi-Fi
The Wi-Fi Alliance runs a certification program through a number of authorised test laboratories (ATLs). Their certification process provides a greater focus on user experience, stressing that they go beyond mere implementation to 'real-world' performance. This involves testing against test-bed devices, which are generally commercially available products, as well as against test equipment. To certify a product, companies need to be regular, affiliate or sponsor members. Unlike the other standards, Wi-Fi does not allow adopters (the lowest membership level) to certify products.

Wi-Fi certification consists of a number of mandatory tests, accompanied by optional tests. The mandatory tests include at least one of the base 802.11abg wireless interfaces and the security elements contained in WPA, WPA2 and EAP. Optional requirements include 802.11n, the country-specific content of 802.11d and 802.11h and the WMM and power-save extensions. Over time, more elements become mandatory, as the market evolves. There is also a joint certification program run with the CTIA [14] to check coexistence of Wi-Fi devices in cellular phones. This is the CWG-RF certification (Cellular Working Group – RF).

In recognition of the emerging market for Internet devices, which may have different user interfaces and use models, the certification process now has an element to certify an application-specific device (ASD). Typical examples of these include medical devices, barcode readers, set-top boxes, VoIP phones and wireless picture frames. A number of ASD test plans have been developed that can be used. Companies are also able to submit new ASDs for consideration if they feel that the published ones do not address their current product. The approval time for a new ASD is around 30 days, after which the certification process can begin.

The certification process allows modules to be qualified, which can then be used without retesting in another certified device. As with other certification programs, there are strict rules prohibiting any changes to hardware, firmware or host drivers. Unlike other standards, there is also a limit to how a module can be reused. The module certification can only be passed to one level of company. If it is incorporated into another product that is integrated into a third product from another manufacturer, then that third manufacturer will need to recertify it. A nominal listing fee is payable when using a pre-certified module. Modular certification is not available for the CWG-RF certification.

Because the upper layer stack is often implemented in a host PC, the behaviour and performance of a Wi-Fi product may differ according to what it is connected to. This has been a particular problem with Windows XP and Vista. As a result, module certifications may need to be tested with a number of different operating systems, and the modular approval will be limited to using the module with these operating systems.

Products that are rebranded by another manufacturer can be certified using the dependent product policy. Again, this can only be devolved once, so rebranding a rebranded product is likely to involve recertification.

10.5.2.3 ZigBee

The ZigBee Alliance allows manufacturers to take a basic platform and add additional profiles and functionality, whilst having the ability to reuse much of the testing that has already been done for that platform.

ZigBee starts off with the underlying 802.15.4 MAC/PHY. It assumes that this is compliant with the 802.15.4 standard, although there are a number of elements of that which are optional for a ZigBee product. Testing of the 802.15.4 layer can either be performed in-house, or contracted to an external test lab. The ZigBee Alliance has published a guidance document for checking compliance of this layer.[15]

On top of the MAC/PHY, the ZigBee certification process covers the NWK and APP layers and has options for manufacturer-specific or ZigBee profiles. For modules, ZigBee certifies up to the top of the APP layer, which it classifies as a ZigBee compliant platform (ZCP). This creates the basis for manufacturers to build full ZigBee products, with the addition of one or more profiles.

For manufacturer-specific profiles, higher-layer tests are essentially limited to coexistence, ensuring that these will not cause problems with any other ZigBee devices operating in the same space.

Full ZigBee application profiles are tested for interoperability against a comprehensive set of test cases. Products that successfully pass certification are termed ZigBee certified devices and can be sold using the ZigBee Alliance logo.

ZigBee grants a free license to universities to use ZigBee IP within their internal projects. If any of the resulting products are commercialised, then they must go through the qualification process.

10.5.2.4 Bluetooth low energy

The qualification for Bluetooth low energy follows the same format as Bluetooth. There is a minor distinction, as the end product listing (EPL) is applied to a device containing a complete solution up to and including GATT and GAP. The addition of GATT-based profiles is covered by a free verification program outside the qualification scheme.

10.6 Open-source protocol stacks

Bluetooth, 802.11 and ZigBee have all attracted open-source software groups, which have developed, or are developing, protocol stacks. These are widely used in university departments, and have been adopted for a number of commercial products.

Designers using an open-source stack need to understand that they will need to take it through the appropriate qualification process. Wireless qualification requires a design to be static. That contrasts with the open-source community, which is constantly updating its implementations. Hence, a manufacturer using an open source stack will need to make the decision to freeze a specific build and then qualify it. As these stacks generally contain everything above the baseband layer, this is a significant task. Any future updates based on a later build will need at least incremental requalification. This cost, which is often missed in an initial analysis, may be greater than the cost of licensing a commercial stack.

There is an academic debate about the validity of using open source with any organisation that requires members to join and sign an IP agreement in order to gain access to use the standard. Such membership requirements conflict with the GNU public license. Whatever the legal merits of this conflict, it has not prevented open-source stacks being used successfully in commercial products.

A number of software stacks and drivers have been written for 802.11 chipsets, often aimed at low-power operation. These can be more difficult to quantify in terms of qualification requirement. If the sole function is to connect to an access point, and no use is made of any of the security procedures developed by the Wi-Fi alliance, i.e., WAP, WPA and WPA2, then any standards-based qualification requirements may not be necessary. If any of these are included, for which the Wi-Fi Alliance holds the IP, then qualification will be required.

This may present problems, particularly for some embedded M2M and low-power applications, where the product may not incorporate enough of the mandatory Wi-Fi Alliance features to allow it

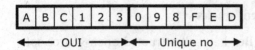

Figure 10.1 Structure of device address

to pass qualification. This is a grey area, where in theory products may infringe IP. If you are designing anything that falls into this category, check the status with the silicon vendor and an IP expert.

10.7 OUI – the device address

One small additional cost many companies will need to incur is the purchase of an organisationally unique identifier or OUI from the IEEE registration body.[16] The cost is currently $1650. This is a unique six-digit hexadecimal reference, which is used as the first half of a wireless device's digital address, whether that is Wi-Fi, ZigBee or Bluetooth. The full address is formed by adding a further six unique characters after the OUI, which increments for each device (Fig. 10.1).

If you are using modules, these will almost certainly have been preprogrammed with the OUI of the module manufacturer. There is no need to change that, but anyone connecting to your product will be able to see who the module manufacturer is from the OUI in the address. For many companies this is not a concern, but purchasing your own OUI means that your products will be identified with your company. Anyone can look up the owner of an OUI at the IEEE registration website.[16]

The six-character serial number following the manufacturer's address must be unique. Most wireless addresses are held in non-volatile memory. If you intend to program you own OUI into a product, then this needs to be included in your production test procedure and you are committing yourself to programming every radio that you produce.

If you already have an OUI, this can be used for your wireless products, but each full address must be unique, even if you manufacture both wired and wireless products. If you produce so many products that you run out of addresses, you will need to acquire an additional OUI.

10.8 Production test

One final item to remember is the need for production test. Testing a radio product adds another dimension to final production, as radio testing is not as simple as plugging in a cable. It is likely to mean that new test gear will need to be designed and manufactured, along with new test regimes.

If the product has a removable antenna, a conducted measurement can be made before the antenna is attached. Alternatively, an RF connector can be used that incorporates a switch, allowing a test lead to take the place of the antenna. These will simplify the test requirements, but will add cost to the product. On the other hand, a functional test has the benefit of checking that the RF components have been correctly fitted.

For many products, the antenna is soldered directly onto the PCB, so over-the-air RF testing is required. Testing may be as simple as checking that the transmit signal is present, or could extend to a full parametric test of the radio performance. In the early days of 2.4 GHz chips, there was sufficient variability between chips that many manufacturers performed a comprehensive RF test and adjusted the radio settings as part of this test. Today, the state of chips has improved to the point where there is little variation in performance, and a simple test that the product is transmitting and receiving packets may be sufficient.

The exception may be where an additional power amplifier is used to increase the power above that obtained from the base chipset. In these cases, there is more scope for variation and a power-level test may be beneficial.

If multiple devices are being tested on the production floor at the same time, this will require the use of screened test enclosures to ensure that measurements are confined to the specific device under test. Test gear of this complexity needs to be planned early in the development phase to ensure that it is ready for first production build.

10.9 References

[1] Federal Communications Commission (FCC), www.fcc.gov.

[2] Certification and Engineering Bureau of Industry Canada, http://strategis.ic.gc.ca.

[3] European Commission for Enterprise and Industry, List of references of harmonised standards. http://ec.europa.eu/enterprise/policies/european-standards/documents/harmonised-standards-legislation/list-references/.

[4] ETSI, Worldclass standards. www.etsi.org/WebSite/Standards/Standard.aspx. Downloadable ETSI standards.

[5] Ministry of Internal Affairs and Communications, Information and communications policy site. www.soumu.go.jp/joho_tsusin/eng/index.html.

[6] European Commission for Enterprise and Industry, Introduction to the R&TTE Directive. http://ec.europa.eu/enterprise/sectors/rtte/regulatory-framework/index_en.htm.

[7] European Commission for Enterprise and Industry, European Commission Enterprise e-services Portal. https://webgate.ec.europa.eu/osn/.

[8] David Seabury, An update on SAR standards and the basic requirements for SAR assessment. www.ets-lindgren.com/pdf/sar_lo.pdf. A good article on SAR.

[9] Continua Health Alliance, www.continuaalliance.org.

[10] US Food and Drug Administration, **How to market your device**. www.fda.gov/MedicalDevices/DeviceRegulationandGuidance/HowtoMarketYourDevice/default.htm. Guidelines on FDA certification.

[11] Bradley Merrill Thompson, Step-by-step: FDA wireless health regulation. http://mobihealthnews.com/4050. FDA certification for mobile devices.

[12] Buffalo, Buffalo settles infringement action by CSIRO. www.buffalotech.com/press/releases/buffalo-settles-infringement-action-by-csiro/. IP settlement between Buffalo and CSIRO over CSIRO's OFDM patents.

[13] CTIA, Bluetooth® compatibility certification program. www.ctia.org/business_resources/certification/index.cfm/AID/11528.

[14] Communications Telecoms Industry Association (CTIA), www.
     ctia.org.

[15] *ZigBee IEEE 802.15.4 PHY & MAC Layer Test Specification.*
     ZigBee document 04319r1.

[16] IEEE Standards Association, Request form for IEEE
     organizationally unique identifier or 'company_id' (aka
     Ethernet address). http://standards.ieee.org/regauth/oui/forms/.
     Registration Authority.

11　Implementation choices

Adding wireless to a product introduces a new set of implementation choices. These have consequences in terms of cost and timescale, which may surprise designers who are used to wired designs. This chapter looks at some of the choices that can be made when adding wireless connectivity to a design and the impact they are likely to have.

In most electronics design it is natural to take the approach of designing with components that are soldered directly to one or more PCBs. Occasionally a module may be used for a specific function, but most designers prefer to design from scratch. Implementing a new wireless design brings in new elements of cost and risk. It is important to understand these before embarking on a wireless design.

11.1　Assessing the options

There is a hierarchy of fairly universal design options across the wireless standards, running from a discrete design all the way to a fully approved module. Each option has an impact on design time, the likely number of iterations to get it right, cost, approvals and production tests. Although there is a correlation between sales volume and minimising cost, other factors, such as time to market, RF expertise and access to design information also come into play in making the choice, particularly if it is a company's first radio design.

11.2　The design architecture

Before talking about the different options, it is worth explaining the available architectural options. For each of the wireless technologies there is a similar three-way split of functional blocks. These are:

- The RF/MAC block, comprising the radio, link manager and baseband/MAC.
- The higher-layer stack. This is often run on a separate processor, although this may reside within the same chip as the RF/MAC. Sometimes it may even run on a different part of the system, as is the case with Wi-Fi or Bluetooth USB adaptors running on a PC.
- The application. In most cases this will run on a separate application processor, although some chips provide a virtual machine, protected application space or supplementary application processor.

Across all of the wireless standards, there are vendors who can provide the individual component parts, or a combination of some or all of them, either in the form of a module or even within a single chip.

Often, the product itself helps to dictate the best approach to take. At one extreme, small low-complexity or low-cost products generally suit the integrated, system-on-chip approach. At the other end, products such as phones or PCs, which have the ability to run stacks within their own operating system, may benefit from the lowest-cost solutions, which only implement the radio and a minimum amount of MAC/baseband, making use of an existing processor to do the rest.

Practical considerations, such as whether the application has already been developed for a specific host, and memory and I/O limitations within chips, will also guide the choice of architecture.

Where single-chip designs are employed, it is important to consider the amount of processing power and program space that will be needed. Wireless standards continue to evolve, both to add functionality and to ensure continuing security for the link. Although it may not be a requirement to upgrade products to add future functionality to a device, it is possible that there will be a need to update firmware to comply with updated security protocols. Enhanced security invariably means more code, so designs that run close to the limit of the available memory capacity should be avoided, unless it is clear that there will be no need to upgrade in the future. The same points apply where a decision is made to consider a ROM-based part for some or all of the wireless stack.

With any wireless design, it is important to remember that the qualification time and cost will be a significant factor in time to market, so the ability to use a pre-qualified design or components may merit greater consideration than would be the case in a non-RF design. Sometimes it is expedient to minimise the risk in the wireless portion of a product.

11.2.1 Chip-based designs

For most designers, the normal inclination is to get hold of the chip data sheet and start from there with a discrete design. This is often the point where a designer discovers that wireless is different. The first indication is the difficulty in getting hold of data sheets.

Despite the fact that all of the standards we are looking at are widely used, the reality is that RF is still difficult. There are still good reasons for it being considered as black magic. Compared with a digital electronics design, it is remarkably easy to get a radio design wrong, particularly at 2.4 GHz. This leads to numerous iterations before it can pass approvals and be shipped. The companies supplying wireless chips have a business model that acknowledges this. It means that they concentrate on three major markets that are the easiest to support:

- High-volume manufacturers, typically mobile phone and PC vendors, who have considerable RF experience within their companies,
- Module vendors and wireless consultancies; companies who have both RF and application expertise and who customise the reference designs or chips to supply other medium volume applications,

- Standard product manufacturers; companies who make standard products, such as wireless mice, headsets and access points, who will work with a reference design from the chip companies and make no significant circuit changes.

The first two of these customers have the in-house knowledge to produce their own RF designs. Companies using reference designs tend to be specialists in high-volume manufacturing for consumer markets. They may vary the housing and add additional external functions, but in general they do not make major alterations to the underlying reference design.

Each of these customers typically buys over a million chipsets a year and will be able to get information and support from the silicon vendor. If you don't fit one of these categories, you are likely to find it difficult to get datasheets, support or even chips.

This is where wireless differs from most other areas. Because wireless is difficult and support is expensive, most chip suppliers will not support companies that do not fit into one of these three categories. It may be possible to buy the chips, but unless a distributor or consulting company can support them, you will be on your own. That means that companies taking this route will need to acquire a thorough knowledge of the standard, RF design and protocol engineering. That may result in a protracted development for the first products, with the possibility of multiple iterations in qualification testing.

11.2.2 Reference designs

As mentioned above, chip companies provide reference designs for the most common applications. If these fit your application, they may be an effective route to market. Depending on your application, the reference design and software toolset that accompanies it may allow you to adjust it to suit your needs.

Most reference designs come as a complete engineering package, along with gerber files for the PCB, firmware and component lists. When taking this route it is vitally important that you do not deviate from these unless you know exactly what you are doing. Many designers have learnt the hard way that changing a component value, or even a PCB pad that is in a critical part of the circuit, can change the design from something that works to something that does not. Reference designs should be taken as immutable – do not change them unless you know what you are doing.

11.2.3 Modules

Modules provide a safe entry into wireless and many companies approaching wireless connectivity for the first time start with a module. Modules offer a number of advantages:

- These are preassembled and pre-tested, so there is little need for RF knowledge and very limited RF production testing is required. In many cases production test may be limited to a functional test only.
- Depending on the wireless standard, modules may be pre-approved, removing the need for standards and regulatory testing either in part, or in its entirety.
- Modules may include higher-layer APIs, which allow them to be more easily interfaced to external circuitry.

There is a wide variety of modules available for all the wireless standards. They range from wireless front-end modules, through complete solutions implemented within single chips to complex modules containing chips for the RF solution and separate application processors that run the protocol stacks and even the user applications. There are also modules targeted at specific applications. These include:

- Bluetooth: headset, serial cable replacement, audio (using A2DP profile), HID (for mice, keyboards and low-latency applications) and medical (using HDP).
- Wi-Fi: MAC/PHY modules for all of the 802.11 variants, low power modules with integrated TCP/IP stacks for automation and low power active RFID modules.

- ZigBee PRO: Modules for endpoint, router and coordinator nodes with fully integrated stacks and application processors.

11.3 Development tools

One of the big advantages of modules is that they let the designer start to evaluate the performance, set-up and integration of the wireless standard. Even if a decision is made to use a chip-based design, the sooner the development team starts to test and understand the wireless link, the better. The best way to accomplish this is to use the development kits supplied by most module manufacturers.

Development kits normally consist of a module attached to a motherboard. Depending on the standard and the complexity of the module, the motherboard may contain anything from a power supply and a set of I/O connectors, to a display, a programmable application processor and a range of switches and connectors.

In some cases, chip vendors also provide development kits. These are ideal for companies aiming to progress to chip-based designs. They may have a steeper learning curve than those that are module-based, as the latter tend to incorporate more complex and developer-friendly application interfaces, where the module manufacturer has done more of the work for general-purpose applications.

Different module manufacturers will support different applications. When selecting the module supplier, it is worth looking to see if there are any with experience in your particular area, as they may have already solved some of the problems that you might encounter.

11.4 Stack integration tools

Where a stack or driver is being integrated onto an external processor, other than on a single chip, then the availability of good integration tools is essential. Where possible, ensure that the stack vendors have experience of running their protocol stack on the target hardware and operating system, as porting a real time stack is not a trivial task.

In the case of Wi-Fi, where there is no defined interface at the MAC layer, drivers are, of necessity, chip-dependent. Although the other standards provide a standard interface, there are still manufacturer-specific commands that can enhance the overall system performance, so it is also important to choose a vendor who has experience with the chipset you are using and who has access to any manufacturer-specific commands. Most of the Wi-Fi chipset suppliers have drivers available for Windows and Mac. They can often also offer Linux source code, which can be ported to other operating systems. If there is no driver for your RTOS and you port a Linux-based one, allow three to six months to develop, test and qualify it.

When choosing a stack, don't forget that stacks should also include good debugging and production test tools. If they are lacking, you will need to write these routines yourself, so this may be a required feature in your choice of stack.

11.5 Deciding on an implementation strategy

There is no hard-and-fast rule regarding which approach to take for a specific product. The following parameters should be considered.

11.5.1 Bill of material cost

This is almost always higher for a module, but is offset by the reduced risk and savings in qualification cost.

11.5.2 Development cost

Normally this is significantly lower for a module, as the design team can concentrate on integration of the wireless functionality as opposed to designing the radio itself.

11.5.3 Integration cost

Although a chip design should, in theory, allow for a more flexible interface to the rest of the product, in practice a module manufacturer will normally have debugged the interface and provided a solid working protocol. So integration of a module is generally simpler. If the interface needs to be optimised, then a chip-based approach may be more appropriate.

11.5.4 RF design

RF design requires a lot of experience. Even with a knowledgeable design team, things can go wrong, requiring multiple iterations. Using a module removes these iterations, giving the dual advantage of reducing design time and hence decreasing development cost.

11.5.5 Approvals

Modules may remove the requirement for approvals altogether (see Chapter 10 for more details), or significantly reduce the time for testing. They should also ensure that the test process is relatively painless. The time and cost for approvals should not be underestimated.

11.5.6 Time to market

The reduction in design time and approvals and regulatory testing can result in a module-based design reaching market three to six months faster than a chip-based design. That can be very important for a first product.

11.5.7 Production test

A module should be pre-tested, so production RF testing can be confined to a functional connection test. The cost of and development time for production RF test equipment is significant.

11.5.8 Size

A module is almost always physically larger than a chip-based design. Because of its physical shape, it may also be inappropriate for small designs, where the circuitry needs to follow the physical form of the device. Conversely, the advanced packaging technology of some vendors who specialise in high volume module design for mobile phones (such as Alps, Delta, Murata and Taiyo Yuden) results in a smaller module than most companies can achieve with a chip-based design.

11.6 Comparison of costs

Every project is different, but over the years I've seen the same costs and development times being repeated across numerous companies for wireless designs. Tables 11.1, 11.2 and 11.3 give an indication of the relative integration costs for a chip- and module-based design.

These obviously vary immensely depending on the available expertise and equipment, and the application. The chip-based design cost is for a first design, where a company is building up a team with the necessary expertise. Subsequent designs should be significantly cheaper. The costs also assume that each design will be right first time and not have to go through multiple spins or approval attempts. Chip-based designs carry a greater risk of re-spins, especially if the design is a company's first attempt at RF.

In practice, designs with total production volumes of fewer than 10000 units will generally benefit from using a module. Above 100000 units, it is likely that a chip-based design will be more economical. Having said which, other considerations may outweigh this simple equation. I have seen perfectly good decisions made using a chip-based approach on a product with a production run in the hundreds as well as modules being used for products shipping tens of millions.

A number of module vendors offer the option of licensing a design to a customer who moves to high volume, so that they can

Table 11.1 *Typical development costs for chip-based designs*

Equipment costs	$100k–250k
(test equipment, software, etc. – one-off costs)	
Hardware development	$75k–125k
(15–25 weeks)	
Software development	$100k–150k
(20–30 weeks)	
Production test equipment	$100k–150k
(development time and hardware)	
Qualification and certification	$50k
Total	**$425k–725k**

Table 11.2 *Typical development costs for integrating modules*

Equipment costs	$20k
(spectrum analyser – one off cost)	
Hardware development	$20k
(4 weeks)	
Software development	$20k
(4 weeks)	
Production test equipment	$20k
(development time and hardware)	
Qualification and certification	$nil
Total	**$80k**

Table 11.3 *Overall wireless project times*

Chip-based design.	32 weeks
Module-based design.	8 weeks

move to placing chips directly onto their own PCB. In practice, most companies redesign their product at this stage rather than confining themselves to a cost-reduction program, but it may be a useful strategy for some.

11.7 Longevity

One question to ask when choosing a wireless supplier is, 'How long will it be available?' Most wireless chip and module suppliers are driven by the market requirements of their largest customers, which can result in their products being available for a limited period.

The life cycle of wireless standards starts off with a wide range of manufacturers designing chips in the hope of becoming one of the leading suppliers. In the course of time that means that around half of the initial suppliers will fail or withdraw from the market. Those remaining will evolve chipsets from their initial, general-purpose chips, (often consisting of a two- or three-chip solution) to a number of generations of single chips. These successful companies usually evolve to supply large volumes to a relatively small number of major customers, who then dictate their roadmaps. This results in a fairly rapid evolution of chipsets to support new features or greater functional integration. As these customers are the overwhelming purchasers of chips, this in turn means that the earlier chipsets rapidly become obsolete.

As an example, since 2001, when the wireless standards started to gain momentum, there have been seven generations of Bluetooth chipset, six generations of 802.11 chipset and three of ZigBee.

The inevitable result of this is that chips become obsolete faster than designers would normally expect. Moreover, because the feature set of chips is driven by customers who demand additional functionality within each new generation, they are rarely pin-compatible with earlier generations. Even when they are, moving to a new chip requires a design to go through all of its radio approvals again.

The lifetime of products within the mobile phone and PC industry is typically 18 months. The underlying phone platforms are used for a couple of generations of handsets, so the life of a wireless chipset can be as little as three years. This is an important consideration for such markets as automotive and medical, where the design time to market may be longer than this, and product life can be in excess of ten years.

Currently these markets do not consume a large enough number of wireless chips to be able to exert much influence over the lifetime of silicon. That is likely to change as wireless becomes more widely deployed, but the long-term availability of a chipset should be determined before embarking on a wireless design.

One way to help insure against this obsolescence is to look at the option of using a module. A number of module manufacturers have a policy of retaining a compatible footprint across successive generations of their modules. They take the most recent chipsets and produce a module that is backwardly compatible with their previous offerings. This has the dual advantage of bringing new functionality to the market, at the same time as preserving compatibility to extend the lifetime of products that use them.

A final point to be aware of, both for chips and modules, is that there is no compatibility between products from different manufacturers. So far, within the wireless realm, there is no example of second source or pin compatibility for wireless chips from different suppliers. In a few cases, module vendors have copied each other's physical layout and pin configuration, but this is still extremely rare. If you find that you need to change supplier for either chip or module, then it will require a full rework of the wireless section of your design. If you take the module route and have space available, it may be prudent to include footprints for alternative modules from different suppliers. However, even in this case, the interface APIs are likely to be different.

In conclusion, make sure you think carefully about the life of the product and whether you have a strategy to cope with changes in the chips or modules that you plan to use.

12 Markets and applications

In the decade since wireless standards emerged into the market, over three billion standards-based wireless chips have been sold and incorporated into products. Despite this huge growth, very many of them remain unused, and where they are deployed, only a few specific applications have emerged.

Intriguingly, the position is different if we look outside the field of wireless standards. If we compare the market for proprietary wireless, then the uses are very many and varied. Proprietary wireless competes directly with standards-based wireless in many areas and dominates in others. Amongst these are wireless mice and keyboards, stereo headsets and remote controls. There are a number of reasons for this and before looking in detail at the market potential for wireless standards, it is instructive to consider why they have not had the expected widespread success.

As an aside, proprietary wireless should not be dismissed as an option for wireless designs. It comes in many forms, is often optimised for a particular application and, as a result, can offer benefits in terms of power consumption and price. It achieves this because it does not come with the baggage that often encumbers a standard.

Where proprietary wireless falls down is evident from its name – it does not offer interoperability. For a product that will never talk to a product from a different manufacturer, proprietary wireless may be the best choice, but that means it is an isolated design. It predicates a decision that a company believes that it can own its own portion of a market in isolation. That can make it very vulnerable to future competition from a wireless standards-based product. What also follows from this is the corollary that the market is not deemed to be enormous.

Although it may not seem intuitive, as volumes increase, so does the power of the single-mode chips associated with standards. Because the same chips can be used across many different markets, they benefit from economies of scale, which can make them very cost-effective. Moreover, the complexities of most standards result in standards-based chips containing significantly more functionality than proprietary chips. Where volumes are great enough, these features can be harnessed to replace external microprocessors and I/O blocks, resulting in a lower overall design cost than using a simpler proprietary wireless part. A good example of this is the Wii controller, which is based around a Bluetooth chip. Although optimised for the Wii, rather than connecting to any other Bluetooth device, it makes use of the functionality of the underlying specification, and gains the cost reductions that come when you ship in excess of 100 million devices.

The demands of interoperability have proved to be something of an Achilles' heel for standards. Rather than enabling a host of different applications, they have confined themselves to a very few high-volume applications. In the case of Bluetooth, that has been headsets for mobile phones. For Wi-Fi it has been access points and Internet access for mobile devices, including phones and laptops. ZigBee is still searching for its break.

As explained at the start of the book, Bluetooth and Wi-Fi got their break with a 'free ride', where they were incorporated into mobile phones and laptops. In the early days, very few users would ever have used this functionality. It is still debatable how many do, but the percentage is growing, although in both cases it is still well below 50% and possibly much lower. What this free ride does is to create an infrastructure of devices that could be connectable. For Bluetooth that means a mobile WAN device that can be used to send data to the Internet. For Wi-Fi it is primarily an access point, providing Internet access.

This is an ecosystem, distinct from cable replacement or local connectivity, where wireless is simply being used to replace a wired link. All of the standards that we have discussed can be, and often are, used purely for cable replacement. In most implementations they emulate a serial port and use a proprietary protocol. So, although they benefit from all of the work that has gone into the standard, they are effectively proprietary in their approach, as devices from different vendors do not talk the same language at the protocol level.

Leveraging the capabilities of a standard is an advantage that should not be ignored. The sheer level of engineering expertise that goes into a standard is vast. All of the wireless standards I've discussed involve hundreds of man-years of effort, usually from some of the leading experts in the industry. They may not result in the simplest implementations, but they are generally well thought out, robust and, most importantly, kept up to date and corrected as issues become apparent. As they evolve, they will encroach on more and more of the proprietary market. They will never replace it completely, but they will increasingly become the first choice of a growing number of design engineers.

12.1 Growing the market

Given the lack of success in expanding beyond a few high-volume applications, it may seem strange to conclude that the role of wireless standards is about to burgeon in other markets. The reason for optimism is that two significant structural changes are occurring in the market.

The first of these is the growing maturity of the wireless marketplace. Up until now, most wireless implementations have been a small part of a product's functionality and, in many cases, were never used. Wireless was incorporated simply because it was a tickbox item for that product's specification. That has led to a fairly static market share amongst chip providers who, having grown rapidly, have begun to reach a plateau of sales. To increase their revenues, they need to find new applications and markets where wireless will be a more fundamental part of the product's identity.

At the same time there are external, political pressures to address a number of global problems. In particular, wireless technology is being looked at to provide solutions for healthcare, energy conservation and management, and improving transport systems. For technology to have an impact in these areas, standards are absolutely necessary. No single company has the resources to roll out a proprietary implementation on a global scale. Instead, the wireless standards are jockeying for position to provide the interoperability vital for mass-scale deployments.

Alongside these, new developments, such as the ability to connect consumer products to the Internet, and to extend a mobile data link beyond a phone to a personal device, open up the possibility of a new consumer electronics market, where individual devices implement Internet connectivity as a key part of their product capability. If fashion-, brand- and technology-conscious consumers take these to heart, it could open up a market for multiple connectable devices attached to every mobile phone. That's a market opportunity that could run into many tens of billions of devices.

To complete the book, I look at some of the initiatives behind these new markets, and the organisations that are helping to drive them, and consider how real the opportunities are.

12.2 Healthcare, wellness, sports and fitness

There is no lack of agreement regarding the need for major change within healthcare systems around the world, as a result of the changes in demographics and the increasing incidence of chronic, long-term disease. Within the developed world, health spending currently varies, but is rising sharply. The OECD figures for 2007 [1] indicate that this ranges from around 6% of gross domestic product (GDP) in Mexico through an average of 10% in Europe to 16% in the USA.

On average, healthcare spending has grown by 20% per decade for the last 40 years – a rate that is no longer sustainable. Continuing at that rate would increase the US healthcare spend to 40% of GDP by 2060. That's without the added demographic pressures of an aging population and the increasing incidence of long-term chronic conditions, both of which will accelerate that growth. Add these, and some analysts believe that the US projections would lead to healthcare consuming 50% of GDP by 2050.

It is an issue that has been largely swept under the carpet and ignored by governments around the world for the last few decades. It has now become a sufficiently urgent priority that political debate is restarting about ways to solve the problem and contain the costs. The diversity of different healthcare provision, from full state funding to private insurance, means that there are many different models, but all face the same issues – trying to keep people out of medical care by becoming more involved in their own health, and reducing the time spent in hospital.

Telehealth, also known as remote healthcare, eHealth, mHealth and a host of other names, is being promoted as the answer. It embraces a wide spectrum, from sports and fitness devices, through preventive and occupational diagnostics, general wellness, management of long-term chronic disease and rehabilitation after surgery, to assisted living.

The biggest hurdle for widespread use is the problem of making these personal medical monitors easy and convenient to use, along with the provision of compelling feedback to encourage people to keep on using them. Wireless removes the inconvenience of cables. This can change consumer medical devices from static variants of clinical devices to portable mobile ones that you carry with you. That's an important step in changing the user perception from the device being a prescription product to a personal purchase.

The key tenets of telecare are the design of simple-to-use, low-cost, non-invasive sensors and personal medical devices, the communications technology to transmit these data and the applications residing on a back-end server. All of these solutions rely on a simple means of transmitting patient data from a sensor to a remote database, where it can be analysed and feedback presented to a patient. By making that data transfer easy, wireless connectivity results in more regular data capture and reporting. As a result, wireless standards are seen as being crucial to providing interoperability at the communications level.

The various initiatives under way place a growing emphasis on self-management of disease and improved prevention. Wireless technology figures in the proposals to cope with this issue as it offers easy-to-use consumer health monitoring devices. The new model of healthcare talks of 'patient empowerment', (note that most medical advocates for change prefer the term 'consumer directed', i.e., they stay in charge), which implies products and services that can be directly purchased in some form by the consumer. A lot of interest has developed around what is dubbed the 'Health 2.0' approach, where web services will provide the means for patients to monitor themselves.

12.2.1 The Continua Health Alliance

A key role in the evolution of the healthcare market is being taken by the Continua Health Alliance.[2] Continua was formed in 2006 to bring together key manufacturers and providers of healthcare with the vision of generating an interoperable ecosystem for electronic health. Members include technology companies, medical device manufacturers and institutions like Britain's NHS and Kaiser Permanente in the USA.

The Continua Health Alliance is generating guidelines for manufacturers, which cover the entire connection from consumer or clinical device, through a WAN interface to the final integration of the medical data into an electronic health record. The aim of the organisation is to build on industry standards to enable a wider ecosystem of connected medical data. By working with all parts of the medical 'food chain', a remarkable degree of common purpose has been achieved.

At the protocol level, Continua products utilise IEEE 20601, with individual devices conforming to the IEEE 11073 device specialisations for their data formats. Currently the guidelines support the use of USB for a wired connection and Bluetooth BR/EDR for wireless connectivity. The next releases of their guidelines will integrate support for ZigBee and Bluetooth low energy.

12.2.2 Health 2.0

The web will have an increasing effect on how we treat health information, and whom we trust with our own health information. We are already seeing the arrival of more interactive sites, often referred to as Health 2.0 sites, which invite a patient to enter personal information to allow them to track their health or the course of a disease. Amongst these, the most prominent are commercial sites like Revolution Health [3] and sites that enable personal or electronic health records (PHR or EHRs) like GoogleHealth [4] and Microsoft's HealthVault.[5]

These sites will grow as consumer devices for measuring blood pressure, heart rate and weight, which can connect directly to the web, become available. As well as recording data, these websites are likely to evolve to provide feedback about a patient's condition. The business models behind these will vary. What they will do is generate an increased amount of medical data, which may well be located outside the preserve of current healthcare suppliers.

This dissemination of patient information will present a major challenge for today's healthcare providers, as it breaks the structure that has been built up by the medical profession over the past few centuries. As such, it will be disruptive – it has the potential to repeat the effect that MP3 and Napster had on music – and could change the whole landscape of healthcare and its ownership. An indication of the pent-up demand is apparent from the fact that by the end of 2009 there were over 2000 medical-related applications available for the iPhone. The availability of wireless-connected consumer medical devices, some of which may implement the Continua Alliance guidelines, will be critical to the growth of this market. Analysts estimate that the new market for health and fitness devices will grow to over 400 million units per year by 2014.[6]

12.2.3 Clinical asset management and lone workers

As well as managing the patient, mobile technology is increasingly being used to track assets within hospitals, whether that is equipment or staff. Wireless LAN technology is being employed in the form of small tags, to log where equipment is located. The benefits are twofold – it's a powerful tool to prevent theft, as an alarm can be generated if equipment is moved outside its expected location – a technique called 'geofencing'. It is equally valuable to track down the closest piece of equipment in an emergency, which can include a member of the medical staff. These monitoring tags are typically small, battery-powered units the size of a box of matches, which run for several years before recharging is required.

Locating staff can be just as important when they're working out in the community as well as within a hospital. A lone worker alarm typically includes a GPS (satnav) system and short-range connection to a mobile phone. When activated by the user pressing a panic button, it makes a network connection that alerts a control centre. The advantage in separating the panic button and the phone is that it allows the button to be less obtrusive, so that its presence does not further exacerbate a difficult situation.

The lone worker alarm is essentially a more mobile version of the fall alarms that are used by over a million people in their homes in the UK. The UK accounts for over 60% of all connected fall alarms deployed worldwide. That statistic illustrates one of the issues faced by telecare systems. Often the cost of deployment is shared by multiple agencies, which have separate funding. Unless there are higher-level mandates to encourage deployment, the inequalities of silo funding can thwart the roll-out of telecare schemes, particularly if they impose a higher cost on one of the participating agencies. Accountants can prove to be a bigger obstacle than the technology.

12.2.4 Assisted living

As the population ages, more needs to be done to allow people to remain living safely in their own home. By 2060 30% of the EU population will be over 65 and the ratio of those to the working population will have more than doubled, from 25% today to 53%.[7] Germany hits the 50% ratio as early as 2035. As these demographics change, it becomes more important not just to keep the older population well, but also to help them to live confidently in their own homes. Assisted-living technologies address this by providing a non-intrusive, monitored environment able to warn carers or family of any issues.

A key part of this is the availability of low-cost wireless sensors that are simple to install and maintain. The latter factor means that they need to run off batteries for several years to reduce maintenance requirements. Both ZigBee and Bluetooth low energy are targeting this market with profiles for assisted-living sensors designed to connect to the Internet. They will allow monitoring services, which may be run by local authorities, independent companies, or even relatives, to ensure that the elderly are living safely. Other monitors, which include basic medical monitors in clothing, can perform the same job using a mobile phone as a link, providing personal monitoring without institutionalising people within their homes. The longer people can remain in their own, familiar environments, the better. Once they are admitted to care or hospital the cost of treatment escalates and their life expectancy decreases.

Each deployment of an assisted-living monitoring system may employ 20 or 30 wireless sensors, including door sensors, carbon monoxide and smoke sensors, and movement and occupancy sensors. Although each element is low in cost, the overall market will grow to several billion devices.

12.2.5 Sports and fitness

The area of health and wellness is a spectrum. An important portion of that is sports and fitness, where wireless is already becoming prominent. Today runners can buy shoes that include wireless sensors that transmit the number of steps to a portable recorder or watch. Wireless heart-rate belts are widely used by runners and most sports manufacturers are experimenting with adding wireless technology to their equipment.

The market is one where consumers are prepared to spend money; to monitor their own performance, to improve their game and, increasingly, to allow them to compare themselves with friends and other competitors. The availability of new, low-power sensors, including accelerometers that can be incorporated into sporting equipment, is leading to rapid innovation. Bluetooth low energy is creating interest from these vendors because of its ability to connect sporting goods to mobile phones, allowing people to interact with a web site, either to record personal performance or to take part in virtual competitions.

As well as personal use, gym equipment is embracing wireless connectivity to record user information. This is helping to drive standardisation to enable consumers to combine their records from inside and outside the gym.

12.3 The telematics and automotive markets

Although the recent global financial crisis has highlighted the overcapacity and inherent weakness of many of the automotive manufacturers, demand for car ownership will continue to rise. As anyone who travels is aware, congestion and journey times are steadily increasing, regardless of the mode of transport.

Governments and municipal authorities have come to realise that building more roads is no longer a viable solution. Instead, technology is being investigated that makes better use of the road systems that we have and ensures that these are an integrated part of a multi-modal transport system.

Alongside this requirement for improved efficiency, there is a desire to reduce the number of fatalities on the road. The level of road deaths is static at around 40 000 per year in both the USA and Europe. After a fall in the 1990s and early part of the twenty-first century, this figure has stabilised. Both the EU and US governments have set targets to reduce the number of deaths and injuries. However, it seems that this will not be achieved by any attempts to change driver behaviour. Instead it will be necessary to use technology to assist the driver or to help manage the car.

An equally important goal of the telematics market is the desire to make vehicles and journeys more efficient, driven by the climate-change agenda. Transport accounts for around 15% of carbon emissions [8] and increasing car ownership in countries like China and India will cause this to grow.[9] It is clear that individuals are not going to stop travelling, so technology is being asked to provide a solution.

12.3.1 Vehicle-to-vehicle communications

Most development effort to ensure safer road journeys and better utilisation of the road network is based on vehicle-to-vehicle communications. The premise is that to improve their standard of driving, drivers need more information available, to make better decisions, and also for the car to be able to act autonomously, to limit the severity of an accident when a driver is unable to control it. As yet, there is little desire for fully autonomous systems, although they are one extension of the current technology trends.

12.3.1.1 Dedicated short-range communications

Dedicated short-range communications (DSRC) [10] has been around as a concept for many years. It goes under a variety of names; currently the fashionable ones are car-to-car (C2C) and vehicle-to-vehicle (V2V). It is a technique for vehicles to use wireless transmission to receive and propagate safety information between vehicles on the road and also between fixed roadside infrastructure elements. It aims to reduce the number of fatalities on the roads and ease congestion by providing active traffic management.

The standard uses 5.8–5.9 GHz radios to communicate current information. It can be used for collision avoidance, collision mitigation (pre-deploying brakes and safety systems), controlling flow with interactive traffic signals and congestion control from condition monitoring and vehicle spacing. Some governments also plan to use it for road tolling. The frequency bands around the world are different, but all reside within the 5.8–5.9 GHz band, so it is straightforward to design products that can be adjusted to work anywhere. Very few road vehicles move between continents, so this is not a major obstacle to deployment.

In the last two years, work on DSRC has gained momentum, as both the EU and the US governments have mandated the roll-out of such systems. Their rationale is twofold. They argue that this level of intelligence in vehicles is required to further reduce the number of deaths on the road. The second strand of the argument is that the technology is needed to increase the capacity of existing roads as usage rises. There was an initial mandate that DSRC should be fitted to all new vehicles in 2011. This will almost certainly slip, but there is pressure to contain any major slippage.

That ongoing pressure has been demonstrated in Europe, where 30 MHz of spectrum was allocated in the 5.9 GHz frequency band in 2008. This allocation was not expected to be made until 2010, but its early adoption has helped galvanise the industry.

12.3.1.2 The DSRC standards

The wireless hardware for DSRC is based on a variant of the 802.11 wireless LAN standard, known as 802.11p. The initial aim of this group was to use the 802.11a MAC, with a new PHY running at 5.8–5.9 MHz in place of the 5.1 GHz 802.11a transceiver. As the use cases have been refined, the MAC has diverged more and more from the 802.11 version, so that it is now a substantially different beast. The main reason for this is that vehicle-to-vehicle connections need to be made and authenticated quickly, as the connection opportunity, whilst moving, may be less than a second. Although existing chips and firmware have been used for initial prototypes, the drivers and MAC have now a sufficient number of new requirements that they need new silicon.

This raises a concern about deployment times. Chip vendors are reluctant to design a new chip until they see a significant market opening up. On the other hand, test and development cycles for automotive equipment can easily be five years or more. This means that there is currently no source of low-cost silicon for these systems. Moreover, it will take a significant deployment of DSRC-enabled vehicles before drivers will see any benefit. Minimum estimates of this are around 10% of vehicles on the road. Until that figure is reached, there is no incentive for DSRC to be installed on the massive stock of legacy vehicles, so chip vendors are reluctant to spend money to design 802.11p chips. Government support may be required for the industry to get over this impasse.

Above the wireless hardware, a new protocol stack is being defined to control the way connections are made, to prioritise the information sent, to decide how that information is arranged and to manage the topology of connections between moving vehicles and the infrastructure around them. The fluid and transitory nature of these connections means that a traditional TCP/IP approach is inappropriate, so a fundamentally new protocol stack is being developed by a number of standards and working groups. Current proposals envisage a network layer that can talk to a variety of radios around the vehicle, and provide data streams for information and time-critical safety applications.

A lot of this development has been academic, funded by major research grants in the USA and Europe. Development has been slower than anticipated and not necessarily focused on the key requirements. Out of this frustration, a group of motor manufacturers set up the Car2Car Consortium,[11] whose aim is to take the relevant parts of existing work, both within 802.11p and higher-layer protocols, and produce an interoperable standard capable of deployment. This has accelerated the development within Europe, with the Car2Car Consortium recently supporting the establishment of an ETSI working group to move towards an international standard.

12.3.2 Vehicle and driver monitoring

The second strand of telematics is improving safety and fuel efficiency by modifying driver behaviour. Whereas DSRC is largely aimed at safety and needs a large number of deployments to show any benefit, vehicle and driver monitoring is suitable for fitting to existing vehicles.

Most of these systems work by monitoring the information available on a vehicle's OBD (on-board diagnostics) port. These have been fitted to most vehicles in the world since 2002. These data can be used to determine the fuel economy, as well as the distance driven and how the vehicle is being driven. The data are processed, either locally or after uploading to a remote server, where they are used to provide feedback to the driver or owner.

These systems are being deployed for a number of applications, including fleet monitoring, pay-as-you-drive insurance and logging the behaviour of young drivers. Wireless is widely used to connect the unit that plugs into the OBD port to an external connection, typically a mobile phone or a Wi-Fi access point. A growing market is also developing in units capable of sending data to applications on the driver's smartphone.

12.4 Smart energy

Smart energy and the smart grid are two of the most popular industry buzzphrases. Smart energy has evolved from an earlier simpler form of automated meter reading (AMR) to a concept encompassing the whole of the energy ecosystem, from power generation and grid infrastructure, down to individual utility meters, the means for a consumer to monitor usage in real-time and the connection and control of domestic appliances.

The original driver was the goal of reducing the cost of reading consumer meters by developing automated meter-reading techniques, typically either by the meter sending data back over its powerline connection, or through a wireless link. That included wide-area cellular links, as well as short-range wireless links, which could be accessed from a van driving past the property. In practice, the cost of these meters limited the deployments to a few areas where other financial factors, such as the level of fraud, justified the installation cost.

What has changed is a growth in concern about energy security over the coming decades, allied to a need to reduce carbon emissions. Throughout much of the world, generating capacity has not kept up with the increase in demand. There is a growing consensus that there may be a potential shortfall in generating capacity from 2015. That is too short a time to build a significant number of new power stations. The situation is made worse by the fact that a number of first-generation nuclear plants are reaching the end of their operating lives and that carbon emission limits may mean that other older fossil-fuel powered plants will either need to be closed, or be expensively retrofitted to curb emissions.

With this prospect, smart energy offers the best hope of limiting the rise in consumer demand. Smart energy meters and appliances can alert users to their consumption, in the hope that they will modify their usage and behaviour. It also provides utilities with the ability to price energy on an hourly basis, so that consumers can be financially nudged to modify that usage by the application of differential pricing throughout the day.

The potential of smart metering could be still more intrusive. By providing usage information back up the network, it allows utilities to get a better understanding of usage patterns. They gain the ability to reduce demand selectively by turning off connected consumer equipment at times when the grid is stressed. How this will chime with consumers has yet to be tested, but the option of losing air conditioning for half an hour is probably more acceptable than losing all power as a result of a brownout. In some parts of the USA, energy shedding schemes like this are already in use.

The key tool to achieve this is the smart meter. The concept of the smart meter has evolved to be considerably more than the old automated meter reading. Tomorrow's smart meter will be capable of a two-way dialogue with the energy supplier, sending usage information back to the utility and updating the consumer with the energy usage within the house. It may include instantaneous pricing, and potentially may interrogate and control all of the major appliances within the home, with the ability to switch them off, or reduce their consumption demand, either from a local consumer energy monitor, or remotely from the utility or another service provider. For a good description of the smart meter concept, see the SRSM report.[12]

The development and deployment of smart meters pose some major challenges. The general consensus is that they need to implement a wireless link, both for ease of installation and the subsequent addition and removal of appliances around the home. Whilst electricity meters have power available to run a wireless or powerline link, gas meters do not. They require a wireless connection to talk to other smart meters. They also impose a requirement that the wireless standard needs to be low power, as battery life for these meters typically needs to be a minimum of 15 years. The wireless range needs to cover an entire house, although there is no reason that all of the connectivity should be wireless. Where electrical appliances are being controlled, smart electricity meters could connect to these through a power line connection. For gas and water meters and sensors, wireless is likely to be a necessity. The other benefit conferred by wireless is the ease of installation of displays and sensors, particularly for wall-mounted sensors where wiring would be costly and unsightly.

Whereas deployment of AMR has stalled because of cost, smart-meter deployment is likely to happen because of government mandates. Governments around the world are legislating for smart meters to be installed, with many planning for a full change-over of meters in the timeframe of 2015 to 2020. Although that may seem a long gestation, the practical effort of replacing every meter is a massive task, which is likely to be difficult to achieve in a much shorter time, especially as most countries are only just embarking on trials.

Initial attempts to design smart meters either used proprietary wireless designs or were based on variations of the standards. The original belief, at the height of enthusiasm for AMR, was that interoperability was not one of the most important issues, as utilities would be responsible for the meter infrastructure. Today that view has changed; as energy suppliers have realised the power of smart metering, they see the need to connect to appliances and displays from other manufacturers, as well as linking multiple utility meters within the home.

Wireless security is a major concern for the smart meter industry. In unregulated markets where multiple meters are linked to a wireless gateway (possibly the electricity meter), each utility requires a security framework that prevents any other supplier to the same property or complex being able to access their readings. Any failure in this would open up the opportunity of enticing customers to move supplier by offering preferential pricing schemes. Without that assurance, there may be a reticence to allow a single meter to be used as a gateway for other utility services. An associated risk is that if smart meters are installed with low levels of security, which allow them to be hacked, the ensuing negative media coverage and consumer concern could put the whole market back by several years.

These key requirements of interoperability with appliances, security of data and ultra-low power are leading to a re-evaluation of the wireless standards. ZigBee currently has the most traction, having developed a specific smart-energy profile to address this market. The potential size of the market is ensuring that other standards are vying for a share, amongst them Bluetooth low energy, Z-wave, Wavenis,[13] and wireless M-bus.[14]

12.4.1 The key opportunities

The overall size of the smart-energy market is immense. Pike research [15] has estimated that more than 250 million smart electricity meters will be installed worldwide by 2015, representing a replacement of over 18% of the global installed base. This equates to an overall revenue of $19.5 billion. The EU has set a goal that 80% of European meters will have been replaced by smart meters by 2020. These figures are only for electricity meters. Adding gas meters, energy displays for each meter and controllers for major appliances moves that figure into the billions of wireless devices.

There will be limited opportunities for new product companies within the smart energy metering market itself. The market for meters is covered by around half a dozen well-established manufacturers, who have long-standing relationships with the utilities. Given that the average life of a meter is around 25 years, and reliability is key, it will be difficult for new companies to break into this particular sector.

However, none of these companies is a wireless expert. That provides an opportunity for specialist companies to partner with them and provide wireless expertise. If history is a guide, those technology providers who are successful are likely to be acquired by the meter manufacturers.

The same applies to appliance manufacturers; wireless expertise is unlikely to generate new companies supplying air-conditioning units or dishwashers. But there are opportunities for companies to provide them with chips, modules and integration expertise.

There is potential for new entrants in smart energy, however, in the monitoring and management devices likely to be required in every home. These range from individual power meters, which may be incorporated into devices or power sockets, through to displays and management controllers, which inform the householders of energy usage and allow them to make decisions about it. In connected homes these may be web-based, being accessed either from a PC or a mobile phone.

There are a few unresolved questions within this market. The first is: who will pay for this massive infrastructure roll-out? Some estimates have put the cost of installing a replacement meter as high as $600.[16] At this level, the utilities are likely to request government support, or pass the cost on to the consumer, both of which may be a difficult proposition to sell. A useful discussion of the possible models is available in a report by Baringa.[17]

The second major question is the choice of standards for these meters. Most of the industry has still not taken on board the need for interoperability. This is a new concept for the industry, as utilities have never interacted with equipment from other companies.

Their thinking has not evolved much further than considering the link between the smart meter and an energy display. To obtain a real benefit, there needs to be one standard that extends across all meters, displays and appliances.

There are good technical arguments for the allocation of a new, reserved, lower frequency band dedicated to smart metering. This would result in lower power consumption, greater range and no problems from interference. It would require government under-standing and action, and there is currently no sign of that. Instead we are likely to see a roll-out of meters using incompatible stand-ards with a subsequent upgrade, replacement or abandonment of the technology five years later.

12.5 Home automation

Home automation is a market that has been just about to happen since the 1950s and we're still waiting. It has always suffered from being a technology pull which consumers see as adding complex-ity to their lives, rather than simplifying it. That may be about to change, in large part due to the deployment of smart meters.

Smart meters will only deliver their promised benefit if consum-ers start to pay attention to the energy they use and modify their energy usage as a result. The carrot behind smart energy is that if they don't, they will be penalised by higher energy tariffs, and potentially by government taxes on 'profligate' energy use.

If this does result in a behavioural change, it implies that con-sumers will start to interact with more home controls and program-ming within their daily life, as they learn to use energy monitors and programmers. If this happens, it makes it much easier to inte-grate other home automation devices into the same ecosystem. The important point here is that it will work best if the same inter-operable wireless standards are used to enable the ecosystem. So the wireless standard that wins the smart meter market stands to benefit from the double whammy of a developing home-automation market.

Whereas smart meters will be rolled out, whether or not the consumer ends up paying for them, any home automation expenditure will be discretionary, so for a mass market to develop it will need to be sold on its benefits. The leading opportunity is therefore likely to be in HVAC (heating, ventilation and air conditioning), not least because this is one of the major users of energy. Control of domestic temperature control is almost certainly going to show the greatest payback for the consumer.

The opportunity is twofold. First, it allows the retrofitting of existing installations, as it is typically only the control circuitry that needs to be replaced. This is likely to require multiple sensors for each installation, particularly if more advanced systems make use of external temperature sensors to allow predictive control.

The second opportunity is wireless lighting. Here the monetary returns are less tangible, particularly as the cost of lighting a house will erode over the next decade with the advent of low-energy bulbs and LEDs. The advantage of wireless is in the ease of installation, with switches that no longer require wiring. These may be battery- or self-powered.

As ease of installation is a prime benefit for wireless light switches, there is a possibility that manufacturers may employ different wireless standards from those chosen for smart energy meters, as there is a conflict in their underlying requirements. That won't affect their operation, but it may make it more difficult to develop a smart home, where all energy-consuming devices can talk to each other. Whether or not that happens will be down to the marketplace.

The third opportunity in home automation is the area of household security. There are already plenty of vendors selling wireless home-security systems based on a variety of proprietary and standard wireless technologies. As they are invariably sold as a complete system, they currently have no need to interoperate, so the wireless chips are chosen primarily for cost and low power. As more standardised wireless becomes part of the home infrastructure, there is a growing justification for migrating home security to the same network. At the most basic level this allows lights to be controlled as part of the preventive side of the home network. The security controller can learn the regular usage pattern and be set to repeat this, with variations if required, whenever the house is empty.

A final benefit of sharing a wireless standard is for the next stage of development, which is to give external control of the home network, allowing home owners to check, for example, whether they really did leave the oven turned on when they went on holiday. Whilst some users will relish this prospect, it is likely that most will consider it too big a complication and ignore it. It may still provide an opportunity for services that do it for you, with the arrival of the web-based concierge. That, however, is probably not going to be a major market in the lifetime of this book.

12.6 Consumer electronics

To date, the most successful wireless technology within this arena has been Bluetooth, mainly in the market for headsets. Where it has been successful in other areas, such as its use in gaming controllers like the Nintendo Wii, it has been as a result of an application-specific chip, designed in conjunction with the vendor.

Although there is no reason to suspect that the Bluetooth headset market will not continue to grow, and possibly be augmented with a market for wireless stereo headsets, these are likely to be gradual expansions of today's market.

The markets offering more interesting growth opportunities are those associated with Internet connected devices and phone fashion accessories.

12.6.1 Internet connected devices

These are products that require a connection to the Internet. They are products where that is a requirement, not an option, as the connection and web service is an integral part of their function. It covers static devices, like web cameras and burglar alarms, that fall into the former category of home automation, medical and fitness monitoring devices and connectivity for existing consumer devices, like set-top boxes.

Some of these may well be served by the recent emergence of low-power Wi-Fi chipsets. Although the concept of using Wi-Fi, and its growing infrastructure, has existed for some time, the cost of implementation has been high; power consumption has mandated a mains power supply and installation has required a degree of technical expertise. There also needs to be a wide enough installed infrastructure of Wi-Fi access points and broadband networks to allow units to be bought and sold without the fear of a significant percentage of returns. In many countries, that has now been achieved, although it is still far from universal. As these barriers are overcome, there is greater scope for manufacturers to innovate.

Although many Internet connected devices will be functional items, such as household M2M and security products, there are also applications for toys and more frivolous applications. One of the forerunners has been the Nabaztag rabbit,[18] which connects to a user's Wi-Fi network and which can be programmed to perform a variety of actions, moving its ears and responding to the user. It is one of the few early entrants in this area that has managed to engage enough users to stay in production. Part of that success has been in growing a community around its users. This example highlights something that is key to all of these products, which is a web or web-services interface that provides a compelling reason for users to be bothered to buy and install the products.

Bluetooth low energy has the potential to enable an even larger number of these products, but using the mobile phone as the web link rather than a Wi-Fi network. The principle remains the same, but benefits from promotions and subsidies from network providers who can offer services around these devices. For the network operator, it provides an exciting new opportunity to derive revenue from a user without the need for them to press a button on their handset.

By 2015, the cost of adding Bluetooth low energy to a product, which will give it instant connectivity, will fall below $1. If compelling web services emerge, this may well make Internet connectivity as pervasive in consumer devices as microprocessors are today.

12.7 Fashion wireless

A new market that will be enabled by Bluetooth low energy is for a range of products that connect to a mobile phone. These will not initially be Internet connected devices, but products that can interact with the handset, either as a subsidiary display, or to control features of the phone.

12.7.1 Tags

The first Bluetooth low energy accessory to ship is likely to be a security tag, which uses Bluetooth low energy's proximity function to act as a security key for a handset. This will provide two functions. It will act as a lock for the phone, so a phone's keyboard is only enabled when the tag is nearby. A second feature is that the handset can be set to ring if the tag moves out of range, alerting users to a phone that has been left behind or stolen. The tags will also implement a 'finder' function, which, when activated, will cause the phone to ring, even if on silent, so that a misplaced phone can be found.

12.7.2 Watches

Another market attracting interest is where wireless connectivity is added to a watch. Sales of watches have been declining for most of the last decade, and the industry sees connectivity to a mobile phone as a way of bringing a new generation back to watch usage.

The primary use model is to act as a remote display and control. When a call arrives, this would allow the user to consult their watch to see who is calling and then use buttons on the watch to accept or reject the call. The same watch interface could also be used as a remote control for a headset, either a Bluetooth headset for mobile calls, or as a remote control for a music player streaming to a wireless stereo headset. These latter applications bring a number of usability issues, as well as requiring interoperability with a range of products, so are unlikely to emerge in volume until there is already a well established infrastructure of connecting products.

12.7.3 Bracelets – the new watches

The wrist could become the most fought-over part of the human body for fashion wireless. Although the watch industry would like to use wireless to rejuvenate itself, it is equally possible that bracelets that communicate with, or control, the phone and other wirelessly enabled accessories, will become a more compelling purchase.

Bracelets could combine the proximity function of tags for access control, as well as providing remote control for phones and other devices. By connecting through a phone to a web service, they also allow button presses to communicate directly with the wearer's social networking sites. In the longer term, they provide an opportunity to incorporate physiological sensors that will turn them into discreet personal health monitors.

12.8 Industrial and automation

The industrial market is a slow one to embrace new technology but, over the past few years, has begun to accept wireless as part of its armoury of technologies. It is now working closely with standards groups to ensure that the latest generation of specifications meets its needs.

Industrial automation and monitoring devices demand reliability, not just in terms of physical reliability, but also in their ability to cope with a noisy wireless environment. Over the past decade, a number of different wireless standards have attempted to penetrate this market, but most have been rejected because of concerns over their ability to withstand interference. The standards currently being developed are either extensions of current industrial standards, like WirelessHART, or are based on either Bluetooth or ZigBee PRO, using their capabilities of frequency hopping or redundant mesh topology, respectively.

This market has developed its own protocols over the years, notably FieldBUS, ISA and ModBUS. It has also been fiercely proprietary, particularly in the interfaces for sensors, where manufacturers have specialised in supplying complete systems, along with an ongoing market in replacement sensors.

As elsewhere, there is a growing demand for interoperability from purchasers, allowing them to interchange sensors from multiple vendors. Although this may not affect the overall size of the market, those vendors who are early in implementing wireless standards may well gain market share.

12.9 Self-powered sensors

We are just at the point where low-power wireless is allowing the construction of self-power sensors. As new generations of chips and energy-harvesting sources appear, this opens up a new market for sensor and switch devices that need no power and can be deployed anywhere. It is one of the longer-term markets and probably needs at least one extra generation of chipsets, whether that is for ZigBee or Bluetooth low energy. When the sleep current of these falls below 1 μA, a whole new set of products becomes possible.

12.10 Privacy concerns

Almost all of the applications mentioned above involve measuring or generating personal data and sending it to a remote application. Whether that is done for fun; as with sports gear and devices like the Nabaztag rabbit; as a result of personal choice, in the case of medical monitoring, which may be linked to insurance or assisted living; through employment, as is the case with commercial driver monitoring; or by government edict, it changes the current status and perception of private information.

This opens up a Pandora's box of privacy concerns. At one extreme, it can be seen as the benevolent face of a caring administration or employer. At the other extreme, it may appear to be an unacceptable intrusion by government and a move towards a Big Brother state. Often the perception depends on how the message and application is sold. To take one example, some companies have successfully introduced driver monitoring as a benefit, which they claim reinforces the professional status of the driver. Others have installed the same equipment covertly and, in doing so, have lost the trust of their workforce. As we look at applications encompassing personal healthcare and travel, the potential pitfalls will be even greater.

We've repeatedly seen that making an application compelling for the end user is vital if it is going to succeed. The treatment of privacy of data is equally important. Even if something is mandated by government legislation, the presentation to the consumer of how the data will be treated is fundamental to its acceptance. Any company using wireless to capture or transmit personal data needs to be keenly aware of the emotions this can arouse, along with the fact that these may be very diverse between different groups and countries. Unless they do so, and recognise that within their marketing and business plans, then wireless may not deliver any of its expected benefits.

12.11 Conclusion

The opportunities for new markets to be grown on the back of personal, short-range wireless have never been greater. Most of the technologies being evangelised by governments around the world: healthcare, smart energy and intelligent transport systems, rely on short-range wireless connectivity. The timing is good, as established wireless standards are mature, and ZigBee and Bluetooth low energy have arrived at a point where their low-power ability is much needed.

The speed at which these markets grow is probably not determined by the wireless element, but by the ecosystems around the products, and most importantly by the ease of use and desirability of the applications they enable. It will be innovation in these fields that determines which is the next area to pass the 100-million-product mark.

I hope that this book may make the incorporation of wireless into products a little easier, so that more time can be given to the applications. There will probably never be another killer wireless application, but a large number of smaller markets may add up to a greater whole. Whatever they may be, the next decade will be an exciting one for all the wireless standards.

12.12 References

[1] Mark Pearson, Disparities in health expenditure across OECD countries, (September 2009), www.oecd.org/dataoecd/5/34/43800977.pdf.

[2] The Continua Health Alliance, www.continuaalliance.org.

[3] Revolution Health, www.revolutionhealth.com.

[4] Google Health, www.google.com/health.

[5] Microsoft HealthVault, www.healthvault.com.

[6] ABI research, Wearable wireless sensors. www.abiresearch.com/research/1004149.

[7] European Commission: eurostat, Population growth projections. http://epp.eurostat.ec.europa.eu/portal/page/portal/population/data/main_tables.

[8] World Resources Institute, World greenhouse gas emissions in 2005, (July 2009), http://pdf.wri.org/working_papers/world_greenhouse_gas_emissions_2005.pdf.

[9] Joint Transport Research Centre, Transport Outlook 2008 – focusing on CO_2 emissions from road vehicles, (May 2008), www.internationaltransportforum.org/jtrc/DiscussionPapers/DP200813.pdf.

[10] IEEE 1609 Working Group, DSRC & P1609 project page. http://vii.path.berkeley.edu/1609_wave/.

[11] Car2Car Consortium, www.car-to-car.org.
[12] Local Commnications Development, Report of the SRSM steering group. http://srsmlocalcomms.wetpaint.com/page/ Report.
[13] The Wavenis Open Standard Alliance, www.wavenis-osa.org.
[14] Wireless M-bus. *Communication system for meters and remote reading of meters.* European standard EN 13757–4:2005.
[15] Pike Research, Smart electrical meters, advanced metering infrastructure, and meter communications: market analysis and forecasts, (November 2009) www.pikeresearch.com/research/ smart-meters.
[16] Another blow for UK smart meter rollout, (20 September 2009), www.smartmeters.com/the-news/637-another-blow-for-uk-smart-meter-rollout.html.
[17] Deparment of Energy and Climate Change, Smart meter roll-out: market model definition & evaluation – a report by Baringa Partners. www.decc.gov.uk/en/content/cms/consultations/smart_metering/smart_metering.aspx.
[18] Nabaztag, The first smart rabbit. www.nabaztag.com.

Glossary of acronyms and abbreviations

Wireless technology seems to spawn more acronyms and abbreviations than most. Although I have tried to explain each at its first instance in the text, this glossary is provided for those occasions when they slip your mind.

Most are pronounced as a straightforward sequence of their letters, so AES is always spoken as the three letters: 'A, E, S'. However, a few have gained pronunciations as words in their own right. Where these are in common usage I have tried to indicate them, so that you can drop them into conversation and make sense of them when you hear them.

Acronyms and abbreviation that are specific to a particular technology are marked by [B], [W] or [Z] for Bluetooth, Wi-Fi or ZigBee, respectively. Most are used more widely across these and other communication standards.

A2DP	Advanced audio distribution profile	[B]
ACK	Acknowledgement (pronounced 'ack')	
ACL	Asynchronous connection oriented	[B]
AES	Advanced encryption standard	
AFH	Adaptive frequency hopping	
AMP	Alternate MAC/PHY (pronounced 'amp')	[B]
AMR	Automated meter reading	
AODV	Ad hoc on-demand distance-vector routing	
API	Application programming interface	
APL	Application layer	[Z]
APS	Application support layer	[Z]

ATT	Attribute protocol (pronounced 'at')	[B]
AVDTP	Audio video data-transport protocol	[B]
BAW	Bulk acoustic wave (filter)	
BER	Bit error rate	
BLOB	Binary long object (pronounced 'blob')	
BR	Basic rate	[B]
BSS	Base service set	[W]
BTT	Broadcast transaction table	[Z]
C2C	Car to car	
CA	Collision avoidance	
CCK	Complementary code keying	
CCM	Counter with cipher block chaining message-authentication code	
CSMA	Carrier-sense multiple access	
CTIA	Cellular Telephone Industries Association	
CTS	Clear to send	
CVSD	Continuous variable-slope delta modulation	
DBPSK	Differential binary phase-shift keying	
DCF	Distribution coordination function	[W]
DECT	Digital enhanced cordless telecommunications (pronounced 'decked')	
DFS	Dynamic frequency selection	
DoS	Denial of service	
DPSK	Differential phase-shift keying	
DQPSK	Differential quadrature phase-shift keying	
DSL	Digital subscriber line	
DSRC	Dedicated short-range communications	
DSSS	Direct sequence spread spectrum	
EAP-TLS	Extensible authentication protocol – transport-layer security	
EDR	Enhanced data rate	[B]
EHR	Electronic health record	
EIRP	Equivalent isotropically radiated power	
EPID	Extended PAN ID	[Z]
EPL	End product listing	[B]
ERP	Extended rate PHY	[W]
eSCO	Extended synchronous connection oriented	[B]

ESS	Extended service set	[W]
FCC	Federal Communications Commission	
FDA	Food and Drugs Administration	
FEC	Forward error correction	
FFD	Full function device	[Z]
FHS	Frequency-hop synchronisation	[B]
FTP	File transfer protocol	
GAP	Generic access profile (pronounced 'gap')	[B]
GATT	Generic attribute profile (pronounced 'gat')	[B]
GFSK	Gaussian frequency-shift keying	
GOEP	Generic object-exchange profile	[B]
HCI	Host–controller interface	[B]
HDP	Health device profile	[B]
HFP	Handsfree profile	[B]
HID	Human interface device	
HVAC	Heating, ventilation and air conditioning (pronounced 'aitch-vac')	
IBSS	Independent base service set	[W]
IEEE	Institute of Electrical and Electronic Engineers (pronounced 'eye-treble-ee')	
IETF	Internet Engineering Task Force	
IP	Internet protocol	
IP	Intellectual property	
ISM	Industrial, scientific and medical (frequency band)	
L2CAP	Logical link control and adaptation protocol (pronounced 'ell-two-cap')	[B]
LNA	Low noise amplifier	
M2M	Machine to machine (pronounced 'em-two-em')	
MAC	Media access controller (pronounced 'mack')	
MCAP	Multi-channel adaptation protocol	[B]
MIMO	Multiple in, multiple out (pronounced 'my-mo')	
MITM	Man in the middle	
MPEG	Moving-picture experts group (pronounced 'em-peg')	

MSDU	MAC service data unit	
MTU	Maximum transmission unit	
NAV	Network allocation vector (pronounced 'nav')	[W]
NFC	Near-field communication	
NIST	National Institute of Standards and Technology (pronounced 'nist')	
NWK	Network layer	[Z]
OBD	On-board diagnostics	
OBEX	Object-exchange protocol (pronounced 'oh-bex')	
OFDM	Orthogonal frequency-division multiplexing	
OOB	Out of band	
OSI	Open systems interconnection (reference model)	
OUI	Organisationally unique identifier	
PA	Power amplifier	
PAL	Protocol adaptation layer (pronounced 'pal')	
PAN	Personal area network (pronounced 'pan')	
PBCC	Packet-binding convolution coding	[W]
PDU	Protocol data unit	
PHR	Personal health record	
PHY	Physical layer	
PLCP	Protocol layer convergence procedure	
PSK	Phase-shift keying	
QoS	Quality of service (pronounced 'kwos')	
R&TTE	Radio and telecommunications terminal equipment	
RAND	Reasonable and non-discriminatory (pronounced 'rand')	
RANDZ	Reasonable and non-discriminatory – zero (pronounced 'rand-zee')	
RF	Radio frequency	
RFCOMM	RF communication protocol (pronouned ar-eff-com)	[B]
RFD	Reduced-function device	[Z]
RFID	Radio frequency identification	

RSSI	Received signal strength indication	
RTOS	Real-time operating system (pronounced ('ar-toss')	
RTS	Ready to send	
SAR	Specific absorption rate (pronounced 'sarr')	
SBC	Sub-band codec	
SCO	Synchronous connection oriented (pronouned scow)	[B]
SDP	Service discovery protocol	[B]
SIG	Special interest group	
SPP	Serial port profile	[B]
SSID	Service set identifier	[W]
SSP	Secure simple pairing	[B]
STK	Short-term key	
TCP/IP	Transport control protocol/ Internet protocol	
TIM	Traffic indication map	[W]
TPC	Transmit power control	
UART	Universal asynchronous receiver and transmitter	
UDP	User datagram protocol	
USB	Universal serial bus	
UUID	Universally unique ID	
UWB	Ultra wideband	
V2V	Vehicle to vehicle	
WEP	Wired-equivalent privacy (pronounced 'wep')	[W]
WMM	Wi-Fi Multimedia	[W]
WPA	Wi-Fi Protected Access	[W]
ZC	ZigBee coordinator	[Z]
ZCL	ZigBee cluster library	[Z]
ZCP	ZigBee compliant platform	[Z]
ZDO	ZigBee device object	[Z]
ZDP	ZigBee device profile	[Z]
ZED	ZigBee endpoint	[Z]
ZR	ZigBee router	[Z]

Index